T5-DGC-176

The Economics of the Green Revolution in Pakistan

Mahmood Hasan Khan

The Praeger Special Studies program—utilizing the most modern and efficient book production techniques and a selective worldwide distribution network—makes available to the academic, government, and business communities significant, timely research in U.S. and international economic, social, and political development.

The Economics of the Green Revolution in Pakistan

PRAEGER SPECIAL STUDIES IN INTERNATIONAL ECONOMICS AND DEVELOPMENT

Praeger Publishers New York Washington London

Library of Congress Cataloging in Publication Data

Khan, Mahmood Hasan.
The economics of the Green Revolution in Pakistan.

(Praeger special studies in international economics and development)
Includes index.
1. Agriculture--Economic aspects--Pakistan.
2. Agricultural innovations--Pakistan. 3. Farms, Size of--Pakistan. 4. Wheat--Pakistan. 5. Rice--Pakistan.
I. Title. II. Title: The Green Revolution in Pakistan.
HD2075.5.Z8K42 1975 338.1'09549'1 75-19796
ISBN 0-275-55680-8

PRAEGER PUBLISHERS
111 Fourth Avenue, New York, N.Y. 10003, U.S.A.

Published in the United States of America in 1975
by Praeger Publishers, Inc.

Printed in the United States of America

Dedicated to

The Canadian Taxpayers and Pakistani Farmers

ACKNOWLEDGMENTS

I wish to express my gratitude to Simon Fraser University for granting me a sabbatical leave and to the Canada Council for the generous financial support that enabled me to complete the fieldwork in Pakistan.

While there I enjoyed the hospitality of the Pakistan Institute of Development Economics at Islamabad and the Applied Economics Research Centre of the University of Karachi, for which I am grateful to the directors, M. L. Qureshi and Professor Ehsan Rashid, respectively.

While conducting the field survey in the Punjab and Sind, I received assistance from M. Mustafa, M. Anwar Javed, M. Moosa, Moazzam Hasan Khan and Jang Bahadur. My work in the villages was made easier by Sardar Amir Akbar Khan, Sardar Amir Sarwar Khan, M. Ghaffar Choudhry, S. M. Saleem, Abdul Wadood Khan, and M. Toaha Qureshi.

In the tabulation and processing of the field data, the valuable help of Dr. Mahfooz Ali made it possible for me to use the computer facilities of the United Bank Ltd. in Karachi. Also, I must express my appreciation for the competent work done on the computer by Ahmed Saeed Siddiqui and his associates. I am grateful to Miss Najma Yasmin for her assistance in the statistical work.

The many sessions with Dr. Mahfooz Ali, Hafeez Pasha, and S. Abu Khalid added considerably to my knowledge of the Pakistani economy in general and of its agriculture in particular. Professor Ehsan Rashid made similar contributions. I am deeply indebted to these friends, but I alone must accept the responsibility for any deficiencies that still remain in the book.

My wife Aiysha and our children, Rummana and Mansoor, patiently endured my demands on them. In Pakistan, my mother and others in the family suffered personal inconveniences. I hope they all feel somewhat compensated by sharing the sense of accomplishment I now have.

CONTENTS

LIST OF TABLES

Table		Page

LIST OF APPENDIX TABLES

LIST OF FIGURES

The Economics of the Green Revolution in Pakistan

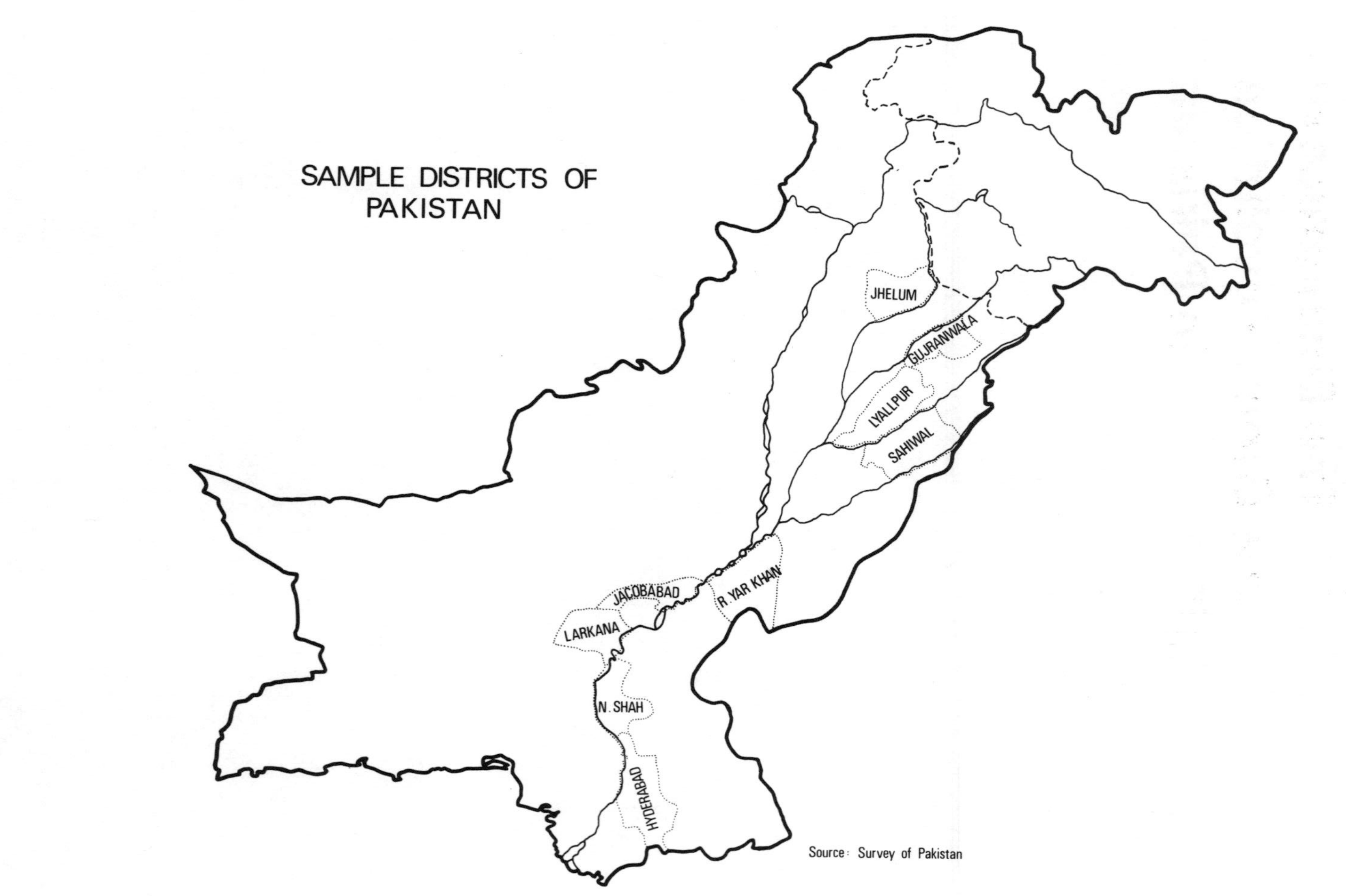
SAMPLE DISTRICTS OF
PAKISTAN
JHELUM
GUJRANWALA
LYALLPUR
SAHIWAL
R. YAR KHAN
JACOBABAD
LARKANA
N. SHAH
HYDERABAD
Source: Survey of Pakistan

CHAPTER

1
INTRODUCTION

In a period when the Third World is caught in what seems to be a grave food crisis, the hopes of a cornucopia created during the middle and late 1960s in countries like Pakistan by rapid adoption and use of the seeds of dwarf wheat and miracle rice (metaphorically called the "Green Revolution") are fast disappearing. Some say that the "Green Revolution" was no more than a brief state of euphoria. In most underdeveloped countries, self-sufficiency in food grains, which many regard as a basic requirement for self-reliance in the development process, remains a dream. In Pakistan, where this goal has been espoused by all governments in the last two decades and where the "Green Revolution" seemed real at one time, widespread hunger (if not mass starvation) is now in the domain of the possible. In 1974 the government was trying to import more than 1 million tons of wheat to meet the minimum consumption requirements of the country.

It was against this background that I undertook this study. It must be said at the outset that no attempt is being made to assess the ability of Pakistan to achieve self-sufficiency in food grains in the near or distant future. It also is not an evaluation of the performance of agriculture in the country. Instead, this study seeks to analyze some specific economic aspects of the "Green Revolution" in the Indus Basin of Pakistan. There are two major objectives:

1. To analyze the adoption and use of the new seeds of wheat (Mexi-Pak) and rice (IRRI), as well as of such inputs as chemical fertilizers, tubewell water, pesticides, and farm machinery
2. To analyze the income and employment effects of the new biological-hydrological-chemical technology.

With a view to achieving these objectives, in 1974 I conducted a survey in nine administrative districts of the Punjab and Sind. The sample districts were selected to represent irrigated and nonirrigated agriculture, the relatively more progressive and more backward agricultural areas, and the areas in which wheat and rice are the dominant crops. Since I was interested in analyzing the interfarm differences, the sample was stratified on the basis of farm size into four groups, to which the group of landless workers was added. The information gathered through a personal questionnaire related to the following:

1. Size, ownership, and fragmentation of farm holdings
2. Cropping pattern, cropping intensity, and factor productivity
3. Time profile of the adoption of the new wheat and rice seeds and other inputs, and the levels used
4. Disposal of crop output
5. Value of crop output, variable cost and net farm income
6. Income from sources other than crop production
7. Size, purpose, and source of agricultural debt
8. Income and employment of landless workers.

The collected data have been used to interpret hypotheses regarded as central to an understanding of the "Green Revolution," which concern the interfarm and interregional differences in the profitability of the new wheat and rice seeds, the benefits from the adoption and use of the new technology, and the impact of new seeds on employment.

Since the analysis in this book is based on cross-section data from farms in Pakistan in 1972-73, I use extreme caution in making generalizations. I hope that the readers will use their judgment as well. The specific limitations of the survey and its results are stated in the text as and when they are required.

This study examines the mechanics of the "Green Revolution" on the farm level, a research area in which very little output is available in the development literature. In Pakistan, certainly one finds no microanalysis of the impact of the new biological-hydrological-chemical technology on farm incomes and employment. (The references to relevant studies are given in Chapter 13.)

CHAPTER

2

METHODOLOGY OF THE FIELD SURVEY IN THE PUNJAB AND SIND

To examine the adoption and use of the new wheat and rice seeds in the Indus Basin, the field survey was conducted only in the Punjab and Sind and only in areas of these two provinces that fulfilled the following conditions:

1. They represent nonirrigated agriculture
2. They represent irrigated agriculture
3. Wheat is dominant
4. Rice is dominant
5. They represent relatively backward and relatively progressive agriculture.

SELECTION OF DISTRICTS

In order to meet the criteria listed above, five administrative districts were selected in the Punjab.

Jhelum is entirely rain-fed and provides a good representation of the nonirrigated districts of the Punjab: Campbellpur, Rawalpindi, and part of Gujrat. Compared with the irrigated districts, agriculture in Jhelum is very backward, in that it is dominated by only two or three low-value crops. The shortage of water combined with a hilly terrain necessitates a limited use of the new inputs.

Gujranwala represents the rice-growing regions of the Punjab: Sheikhupura, Sialkot, and Lahore. It has the largest rice acreage in the Punjab. In the use of new technology, although IRRI varieties do not claim as high a share in the acreage as Basmati (long grain aromated rice), Gujranwala can be regarded as a relatively progressive agricultural region.

Sahiwal, among the irrigated areas of the Punjab, has agriculture characterized by a large number of crops: wheat, cotton, rice, sugarcane, maize (corn), fodder, oilseeds. In the use of new inputs, Sahiwal is one of the leading districts.

Lyallpur may not compare well with Sahiwal in the number of crops grown, but it is certainly a progressive agricultural area in the Punjab. Its agricultural acreage is dominated by wheat, cotton, and sugarcane.

Rahimyar Khan, the southernmost district of the Punjab, represents a relatively backward irrigated area for wheat, sugarcane, and cotton.

In Sind, which is mostly irrigated by the canal system of the Indus, four districts were selected.

Jacobabad receives water from the perennial and nonperennial canals and represents relatively backward agriculture. Wheat and gram dominate the acreage in winter (Rabi) and rice in summer (Kharif).

Larkana is a good representative of the rice-growing area of Sind: Tatta and Sukkur. Perhaps in no other area have the IRRI varieties been adopted to such an extent.

Nawabshah represents progressive agriculture and grows mainly wheat, cotton, and sugarcane.

Hyderabad is a good representative of progressive agriculture, and its acreasge is dominated by wheat, rice, cotton, and sugarcane.

SAMPLE SIZE AND DESIGN

Considering the constraints on time and resources, it was decided that the total number of persons to be interviewed would be 1,000.[1]

The steps followed in determining the design of the sample for the survey are presented below.

Categorization of Respondents

Since the farm holdings in the Punjab and Sind do not have a uniform size, nor do they represent a uniform mode of ownership, it was decided to divide them by size. The sample also included landless workers. In order to meet these requirements, the sample of 1,000 respondents was stratified as follows:[2]

1. Under 12.50 acres
2. 12.50 acres to 25.00 acres

3. 25.00 acres to 50.00 acres
4. Over 50.00 acres
5. Landless workers.

Since no statistics were available for the total Punjab and Sind populations in these categories, except for the data from the 1960 Census of Agriculture (which were both incomplete and inaccurate for this study in 1974), it was decided that, from each of the five categories, an equal number of respondents would be interviewed. With five categories and a total of 1,000 respondents, the number of persons from each category in a village was established at two. Thus, in each selected village in the sample districts, there would be ten respondents.

Distribution of Respondents between the Punjab and Sind

In determining the allocation of 1,000 respondents in the survey districts of the Punjab and Sind, it was decided to establish the percentage share of each province in the total acreage of wheat, rice, cotton, sugarcane, and maize. On this basis, Punjab had 70 percent (700 respondents) and Sind 30 percent (300 respondents).

Distribution of Villages in Each Survey District

The number of villages in each selected district was determined on the basis of the percentage share of each district in the provincial acreage of wheat, rice, cotton, sugarcane, and maize.

	Percent	Number of Villages
Punjab		
Jhelum	6.0	4
Gujranwala	18.6	14
Sahiwal	32.4	23
Lyallpur	27.2	19
Rahimyar Khan	15.5	10
Sind		
Jacobabad	20.4	6
Larkana	20.4	6
Nawabshah	23.9	7
Hyderabad	35.1	11

Number of Respondents in Each Survey District

After determining the number of villages in each selected district, and knowing that in each selected village ten respondents had to be interviewed, the number of respondents were as follows:

	Number of Respondents
Punjab	
Jhelum	40
Gujranwala	140
Sahiwal	230
Lyallpur	190
Rahimyar Khan	100
Sind	
Jacobabad	60
Larkana	60
Nawabshah	70
Hyderabad	110

Selection of Villages in Each Survey District

Having determined the number of villages, and with the village numbers and names in each selected district available from the revenue and census records, the villages were chosen with the help of random tables. The names and location of villages in which respondents were to be interviewed are given in Appendix B.

Selection of Respondents in Each Survey Village

Ideally, the ten respondents should have been selected completely randomly. However, with the problems of records and officials (the records on the basis of farm holding were not complete and the officials were not easily available), it was decided to interview any two persons in each category who were easily available. To reduce the error or bias in responses, no two respondents in a category were to be related.

ENUMERATION OF DATA

To enumerate the data required from the respondents, a questionnaire was prepared. (A summarized version is given in Appendix B.) Only those questions that appeared relevant to the objectives of the study and were communicable to the respondents were asked. To assure the accuracy of the questions, their sequence, and their consistency, a pretesting was done in the district of Sahiwal and was used to modify the questionnaire.

Most questions related to the summer of 1972 and winter of 1972-73. The field enumeration was done during the spring of 1974. The enumerators were given verbal and written instructions for conducting the interviews.

EDITING OF DATA

After the questionnaires had been filled in, they were arranged by farm category for each district. Each was numbered and coded for the relevant farm category and district. Since the tabulation of data was to be done both manually and by computer, I edited the raw data to suit the requirements of tabulations. For some questionnaires, where the information collected apparently was either incorrect or illegible, personal judgment was used.

NOTES

1. For a good discussion of surveys, sample size, and so on, see E. Mueller, "Generating Micro-Data in Less Developed Countries Through Surveys: Some Experience in Asia," in E. B. Ayal, ed., Micro Aspects of Development (New York: Praeger, 1973), pp. 101-17.

2. The basic reason for stratifying the farm holdings into four groups is that it helps to identify the "small" and "large" holdings. Farm holdings of under 25 acres are regarded as "small" and those of over 50 acres as "large." The group of 25-50 acres should be regarded as "intermediate" farms.

CHAPTER

3

SIZE, FRAGMENTATION, AND OWNERSHIP OF FARM HOLDINGS

AVERAGE SIZE OF FARM HOLDINGS

The average size of farm, as shown in Table 3.1, is larger in Sind (34.8 acres) than in the Punjab (29.5 acres). For each farm size the same picture emerges, except in the case of 50 acres and over, where, because of the inclusion of one farm of 700 acres in Rahimyar Khan, the average is lower in Sind. Also, within each category the difference between the upper and lower limits is greater in Sind.

Considering the two provinces separately, there is a great divergence in the average size of farm holding between the sample districts of the Punjab: Jhelum is at the lower limit and Rahimyar Khan at the upper. Rahimyar Khan reports the greatest size because of one very large farm. In Sind, the average farm size does not show a great divergence between the sample districts: the range between the smallest and largest size (in Larkana and Hyderabad) is about 33.0 to 36.0 acres. Similarly, within each category the inter-district disparity is smaller in Sind.

The greater average size of farm holding in Sind has some significance in explaining the difference in farm income between the Punjab and Sind.

FRAGMENTATION OF FARM HOLDINGS

While the size of farm holding in Sind is larger, there is also a greater degree of fragmentation. As shown in Table 3.2, over 68 percent of the holdings in Sind and 40 percent in the Punjab are fragmented. The average number of fragments in Sind is 4.1 against

TABLE 3.1

Average Size of Farm Holding, in Acres

District/ Province	Average Size					Upper Limit					Lower Limit				
	Under 12.50	12.50 to 25.00	25.00 to 50.00	Over 50.00	Average	Under 12.50	12.50 to 25.00	25.00 to 50.00	Over 50.00	Average	Under 12.50	12.50 to 25.00	25.00 to 50.00	Over 50.00	Average
Jhelum	5.0	17.6	31.3	60.0	13.9	10.0	22.5	38.0	70.0	35.1	2.0	13.0	26.0	50.0	22.8
Gujranwala	5.6	18.9	36.3	54.8	24.8	7.8	24.0	45.0	70.0	36.7	3.5	12.5	25.0	50.0	22.8
Sahiwal	8.1	17.7	32.0	64.8	30.7	12.5	25.0	50.0	100.0	46.9	5.0	12.5	25.0	50.0	23.1
Lyallpur	8.1	15.8	30.0	64.4	28.7	12.0	23.0	40.0	95.0	42.5	5.0	12.5	25.0	50.0	23.1
Rahimyar Khan	8.8	18.4	31.4	117.9	38.1	12.5	25.0	50.0	700.0	196.9	3.0	12.5	25.0	50.0	22.6
Punjab	7.5	17.5	32.2	70.4	29.5	11.1	24.0	44.6	207.0	71.6	3.7	12.6	25.2	50.0	22.9
Jacobabad	9.3	19.0	35.3	77.0	35.7	12.0	24.0	50.0	140.0	56.5	6.0	13.0	26.0	54.5	24.9
Larkana	8.0	19.3	36.6	64.9	32.7	12.0	24.5	50.0	90.0	44.1	3.3	13.0	27.1	47.0	22.6
Nawabshah	7.9	17.3	36.0	75.0	34.0	12.0	22.0	45.3	101.0	45.3	3.0	14.0	27.0	54.0	24.5
Hyderabad	9.7	19.3	35.6	84.1	36.1	12.5	24.0	50.0	115.0	50.4	4.0	13.0	28.0	50.0	23.8
Sind	8.9	18.7	35.8	76.5	34.8	12.1	23.6	48.8	111.5	49.1	4.1	13.3	27.0	51.4	27.9

Source: Compiled by the author.

TABLE 3.2

Distribution of Farm Ownership and Fragmentation

District/ Farm Size	Percentage Distribution by Ownership				Fragmentation	
	Owners					
	Individually Owned	Collectively Owned	Tenants	Lessees	Percentage Fragmented	Average Number of Fragments
Jhelum	85.0	10.0	30.0		85.0	7.5
< 12.50	75.0	12.5	25.0		75.0	3.8
12.50-25.00	100.0		28.6		85.7	5.7
25.00-50.00	66.7	33.3	66.7		100.0	5.0
> 50.00	100.0				100.0	21.5
Gujranwala	84.8		15.2	10.9	23.9	2.3
< 12.50	89.2		14.3		96.4	2.0
12.50-25.00	95.8		4.2	8.3	83.3	2.3
25.00-50.00	78.6		21.4	14.3	35.7	2.3
> 50.00	66.7		25.0	33.3	58.3	2.3
Sahiwal	91.3	8.2	4.3	15.8	56.5	2.4
< 12.50	100.0		4.3		15.2	2.0
12.50-25.00	95.9	2.0	10.2	14.3	65.3	2.3
25.00-50.00	83.7	16.3	2.3	23.3	65.1	2.3
> 50.00	84.8	15.2		26.1	80.4	2.7
Lyallpur	79.1	12.2	8.8	8.8	16.9	2.3
< 12.50	94.7		5.3			
12.50-25.00	73.7	7.9	29.0	2.6	5.3	2.0
25.00-50.00	92.1	44.1		10.5	13.2	2.0
> 50.00	52.9			23.5	52.9	2.4
Rahimyar Khan	77.0	13.5	6.8	6.8	50.0	2.9
< 12.50	53.3	46.7			93.3	3.2
12.50-25.00	90.0	5.0	5.0		45.0	2.7
25.00-50.00	80.0	5.0	10.0	10.0	50.0	3.0
> 50.00	70.0	5.0	10.0	15.0	20.0	2.0
Punjab	84.4	5.2	8.9	11.0	39.6	2.9
Jacobabad	68.1		29.8	2.1	87.2	7.5
< 12.50	100.0				100.0	2.9
12.50-25.00	83.3		16.7		91.7	6.2
25.00-50.00	41.7		58.3		66.7	8.3
> 50.00	50.0		41.7	8.3	91.7	12.5
Larkana	50.0	16.7	31.3	10.4	62.5	3.1
< 12.50	91.7		8.3		16.7	2.5
12.50-25.00	75.0	8.3	16.7		66.7	2.1
25.00-50.00	25.0	33.3	41.7		75.0	3.0
> 50.00	8.3	25.0	58.3	41.7	91.7	4.0
Nawabshah	48.2	21.4	30.4		75.0	3.0
< 12.50	100.0				21.4	2.0
12.50-25.00	57.1	28.6	14.3		92.9	2.2
25.00-50.00	28.6	28.6	42.9		85.7	3.0
> 50.00	7.1	28.6	64.3		100.0	3.9
Hyderabad	29.8	35.7	31.0	4.8	56.0	3.0
< 12.50	13.6		86.4		45.5	2.3
12.50-25.00	40.9	31.8	22.7	4.5	50.0	2.4
25.00-50.00	45.0	45.0	10.0		55.0	3.4
> 50.00	20.0	70.0		10.0	75.0	3.5
Sind	46.2	21.4	30.8	4.3	68.4	4.1

Source: Compiled by the author.

2.9 in the Punjab. In the relatively more backward districts (Jhelum and Jacobabad), a higher percentage of farm holdings is fragmented and the number of fragments is the highest. One interesting feature is that in the Punjab the number of fragments is not correlated with farm size except in Jhelum, while in Sind the number of fragments almost always increases with the size of holding. There is no discernible trend in the percentage of fragmented holdings as the farm size increases. However, in Sind a higher percentage of the holdings is fragmented in almost every farm category.

OWNERSHIP OF FARM HOLDINGS

As shown in Table 3.2, over 84 percent of the holdings in the Punjab are owned individually. In Sind, on the other hand, the tenancy system is more pronounced: almost 31 percent of the holdings are held by tenants, as against only about 9 percent in the Punjab. Also, in the Punjab, 11 percent of the holdings are on lease, but just over 4 percent in Sind.

In the Punjab there is no large difference among districts in the percentage of individually owned holdings; but in Sind, the more progressive districts of Nawabshah and Hyderabad have a high percentage of collectively owned holdings. In the Punjab, Jhelum has a high percentage of holdings cultivated by tenants.

From the preceding it is clear that in Sind the average size of holding, the extent of fragmentation, the number of fragments, and the extent of tenancy are all greater than in the Punjab. These results are quite consistent with the 1960 agricultural census data and with the general impression of researchers that Sind has a more bimodal agricultural structure.

CHAPTER

4

CROPPING PATTERNS, CROPPING INTENSITY, AND FACTOR PRODUCTIVITY

Considering the variety of climatic conditions in the sample districts of the Punjab and Sind, the cropping patterns, cropping intensity, and factor (land and labor) productivity differ greatly between the areas surveyed. To some extent these differences also are explained by the state of technology and by the amounts of factor inputs used on farms in the sample.

CROPPING PATTERNS

The differences in the cropping patterns reflect not only the variety of soils and climate but also tradition, technology, and the economic attitudes of the farmers. Even within the same district, there are considerable differences as the size of holding changes.

As can be seen in Table 4.1, the share of major crops in the total cultivated acreage is slightly higher in the Punjab (79 percent) than in Sind (77 percent). In the Punjab, wheat (in which Mexi-Pak is the dominant variety) claims 33 percent of the cultivated area; in Sind, rice varieties have about the same share. Rice in the Punjab (mainly Basmati rice) occupies 14 percent of the area, and wheat in Sind claims 23 percent of the cultivated area. In the Punjab, the share of cotton is higher (21 percent) than in Sind (12 percent). In both provinces, however, improved cotton dominates. The share of sugarcane in the Punjab and Sind is more or less the same: 8 and 9 percent, respectively. Maize in the Punjab has a share of 3 percent, but is negligible in Sind.

As the size of farm increases, the share of each major crop is affected differently. In the Punjab, the shares of wheat, cotton, and sugarcane increase with the size of farm. In Sind, while the

TABLE 4.1

Percentage Share of Selected Crops in Total Sown Area

District/ Farm Size	Percentage of Total Sown Area					
	Wheat	Rice	Cotton	Sugarcane	Maize	Total
Jhelum	52.9				0.1	53.0
< 12.50	46.6					46.6
12.50-25.00	48.7				0.1	48.8
25.00-50.00	59.0				0.3	59.3
> 50.00	55.6					55.6
Gujranwala	39.9	43.7	0.2	0.8	0.3	84.9
< 12.50	33.7	41.5		1.6		76.8
12.50-25.00	32.9	39.9	0.1	0.9	0.1	73.9
25.00-50.00	38.9	44.3	0.5	0.9	0.6	85.1
> 50.00	47.2	45.0		0.5		92.7
Sahiwal	29.6	13.9	20.7	5.7	5.5	75.4
< 12.50	28.2	13.0	19.0	8.6	4.2	73.0
12.50-25.00	27.5	13.3	18.4	7.5	6.8	73.5
25.00-50.00	28.9	13.4	18.5	6.1	5.6	72.5
> 50.00	30.8	14.6	22.7	4.7	5.2	77.9
Lyallpur	34.9	0.9	26.8	13.9	3.9	80.4
< 12.50	32.7	0.4	25.7	8.8	5.6	73.1
12.50-25.00	32.9	1.1	27.3	10.8	5.6	77.6
25.00-50.00	33.0	0.7	27.4	14.3	4.2	79.6
> 50.00	36.5	1.0	26.5	14.9	3.2	82.1
Rahimyar Khan	29.2	1.1	37.1	11.8	0.2	79.4
< 12.50	34.9		33.4	7.3		75.6
12.50-25.00	33.2	0.4	34.7	10.4	0.5	79.3
25.00-50.00	31.8	1.1	34.8	12.1	0.2	80.1
> 50.00	26.3	1.5	39.1	12.7	0.0	79.5
Punjab	33.4	13.5	20.5	8.0	3.4	78.7
Jacobabad	14.6	50.1	1.8			66.5
< 12.50	14.2	50.0				64.2
12.50-25.00	12.0	49.9				61.9
25.00-50.00	12.3	51.3				63.6
> 50.00	16.3	49.6	3.3			69.2
Larkana	23.1	49.7				72.9
< 12.50	30.6	50.0				80.6
12.50-25.00	21.7	50.5				72.2
25.00-50.00	24.5	50.0				74.2
> 50.00	22.0	49.3				71.3
Nawabshah	28.7		30.0	29.5	0.4	88.6
< 12.50	42.1		35.8	5.4		83.2
12.50-25.00	27.7		27.4	30.8		85.9
25.00-50.00	30.7		31.4	31.1		93.3
> 50.00	26.0		29.1	31.9	0.7	87.7
Hyderabad	29.6	26.0	20.3	7.5	0.0	83.4
< 12.50	34.1	29.6	17.5	4.6	0.5	86.3
12.50-25.00	27.6	30.0	19.5	6.6		83.6
25.00-50.00	29.3	30.9	15.3	11.2		86.7
> 50.00	29.6	22.6	22.8	6.8		81.7
Sind	23.4	32.9	12.1	8.6	0.1	77.1

Note: The total in the last column is the percentage share of these crops in the total sown area.

Blank spaces indicate data not applicable.

Source: Compiled by the author.

share of wheat generally decreases, with the increased size of farm, the share of sugarcane increases. The share of rice remains unchanged with the increased size of farm. Only in Gujranwala, Sahiwal, and Lyallpur (all in the Punjab) do the shares of major crops increase collectively with the size of farm.

The collective share of major crops is about 53 percent, in Jhelum, with wheat dominating. The rest of the area is occupied by Rabi and Kharif fodder. The share of wheat increases with the size of farm.

In Gujranwala, where the major crops have the highest percentage in the Punjab (85 percent), the share of minor crops is negligible. Wheat (40 percent) and rice (44 percent) dominate the cropping pattern. There is some rise in the acreage of wheat and rice as farm size expands. Certainly less fodder (Rabi and Kharif) is grown as the size of farm increases.

In Sahiwal, where the major crops collectively claim 75 percent of the total cropped area, wheat is the dominant crop (30 percent). Then come cotton (21 percent), rice (14 percent), sugarcane (6 percent), and maize (6 percent). Both cotton and rice areas expand with the increased farm size.

The major crops occupy 80 percent of the cropped area in Lyallpur. Wheat leads (35 percent), followed by cotton (27 percent), sugarcane (14 percent), and maize (4 percent). There is a positive effect of increased size of farm on the acreage of wheat and sugarcane.

In Rahimyar Khan, where the collective share of the major crops is 79 percent, cotton leads (37 percent), followed by wheat (29 percent), sugarcane (12 percent), and rice (1 percent). The shares of cotton, sugarcane and rice increase with the size of farm; in wheat the share declines.

The highest collective share of major crops in Sind is reported in Nawabshah (89 percent), followed by Hyderabad (84 percent), Larkana (73 percent), and Jacobabad (67 percent). In Jacobabad and Nawabshah the share of major crops generally rises with increased farm size, while in Larkana and Hyderabad it declines.

In Jacobabad, rice constitutes 50 percent of the area, with wheat 15 percent and cotton 2 percent. While the share of rice remains relatively stable with increased farm size, the share of wheat tends to increase.

Rice also dominates in Larkana, with 50 percent of the cropped acreage; wheat has 23 percent. The share of rice in this district remains stable with the increased farm size, but the share of wheat decreases.

In Nawabshah, cotton, sugarcane and wheat claim 30, 30, and 29 percent of the cropped acreage. The share of wheat declines with increased farm size, but that of sugarcane increases. The share of cotton varies.

The share of wheat is highest (30 percent) in Hyderabad, followed by rice (26 percent), cotton (20 percent), and sugarcane (8 percent). The shares of both rice and wheat decline with increased farm size, but sugarcane rises.

There are some interesting features of the changes in cropping pattern. Where both local and Mexi-Pak varieties of wheat are sown, the share of Mexi-Pak increases with the size of farm. In the case of rice, mainly the Basmati and IRRI varieties are grown. In the sugarcane districts the share of this crop increases with farm size during the Kharif season. What it all adds up to is that as the size of holding increases, the farmer shifts to cash and high-yield food crops.

CROPPING INTENSITY

Cropping intensity on a farm reflects the extent to which the available cultivated acreage is planted in a season. It is an index of single- or double-cropping. On examination of the data on intensity in Table 4.2, it is clear that the cropping intensity is lower in the Punjab than Sind. Also, it should be noted that in Sind, the intensity falls with increased farm size. In the Punjab, even where there is such a trend, it is not so pronounced.

Looking at the figures of cropping intensity in the Punjab, the two rather backward districts, Jhelum and Rahimyar Khan, report a low intensity. It is only in Gujranwala and Lyallpur that intensity of cropping increases with the farm size.

In Sind, the figures for Jacobabad do not appear to reflect reality, since no independent evidence corroborates the high intensity figures yielded by the survey. The figures for Larkana also appear unrealistic, but they could be explained by the existing cropping pattern in that district.

FACTOR PRODUCTIVITY

To examine land and labor productivity, the yield per acre (which reflects the productivity of land) and the yield per man-day (which reflects the productivity of family and hired labor) have been calculated for the four farm sizes in each district. Although the figures in Tables 4.3 and 4.4 are related to the average land and labor products of all major crops, for this study the yield data of wheat and rice are of principal interest.

TABLE 4.2

Cropping Intensity

District/ Province	Under 12.50 Acres	12.50- 25.00 Acres	25.00- 50.00 Acres	Over 50.00 Acres	District Average
Jhelum	129.2	81.8	92.4	93.4	79.3
Gujranwala	164.2	169.7	174.7	177.1	137.2
Sahiwal	168.2	162.7	164.3	151.3	129.3
Lyallpur	135.2	142.9	153.7	175.3	121.4
Rahimyar Khan	108.5	102.5	100.8	79.6	78.3
Punjab	141.1	131.9	137.2	135.4	109.1
Jacobabad	200.0	200.0	194.3	205.1	199.9
Larkana	200.0	198.5	199.5	186.5	196.1
Nawabshah	191.9	146.1	137.4	138.0	153.3
Hyderabad	101.0	92.3	82.4	95.8	92.9
Sind	173.2	159.2	153.4	156.3	160.6

Source: Compiled by the author.

The first point to be noted in these tables is that in districts where different varieties of wheat and rice are grown, the land and labor productivities of Mexi-Pak wheat and IRRI rice are generally higher than of their local counterparts. It also should be noted that in the more backward agricultural districts (Jhelum, Rahimyar Khan, and Jacobabad), the factor productivities are lower for both local and Mexi-Pak wheat. In Jhelum and Jacobabad, where water is the major constraint, the yields of Mexi-Pak wheat are lower than of local wheat.

The yields per acre and per man-day of all varieties of wheat and rice are higher in the Punjab than Sind. However, the yield differentials between the new and local seeds of wheat and rice vary between the two provinces. For example, the differential of yield per acre between the new and local seeds of wheat is 130 percent in Sind and 82 percent in the Punjab. But in the case of rice, this differential is 61 percent in the Punjab and 48 percent in Sind. As for the differentials of yield per man-day of Mexi-Pak and local varieties of wheat, the new seeds are significantly more productive (432 percent) in the Punjab and less productive in the province of Sind (-28 percent). The new rice seeds are more productive than the local

TABLE 4.3

Yield per Acre, Selected Crops
(in maunds, 82.28 lbs.)

District/	Wheat		Rice			Cotton		Sugar-	Maize	
Farm Size	Local	Mexi-Pak	Local	Basmati	IRRI	Local	Improved	cane	Local	Improved
Jhelum	5.8	5.2							35.5	
<12.50	5.7									
12.50-25.00	4.3								38.4	
25.00-50.00	5.6								34.0	
>50.00	10.0	5.2								
Gujranwala	14.1	19.7	21.1	19.6	30.8	13.8		33.5	10.0	
<12.50	10.0	14.9	21.5	22.2	33.7			42.0		
12.50-25.00	10.0	19.2	20.0	16.2	26.7	7.0		33.0	10.0	
25.00-50.00	15.0	19.2	19.7	19.5	29.1	14.7		32.3		
>50.00	15.8	21.5	23.8	23.4	33.4			31.1		
Sahiwal	17.2	23.9	20.0	22.1	32.7	9.2	14.5	32.7	10.0	18.8
<12.50	17.8	21.0	23.2	21.3	30.0	9.0	13.7	32.9	10.0	19.1
12.50-25.00	17.5	23.3	20.8	21.1	32.8	8.7	15.0	32.0		18.5
25.00-50.00	17.0	24.0	20.4	22.0	31.1	8.8	14.6	34.6		17.9
>50.0	16.9	24.3	18.7	22.5	34.4	10.5	14.5	32.0		19.3
Lyallpur		25.6	19.2	20.0			17.8	46.1	10.7	30.6
<12.50		21.5	20.0	17.3			11.9	34.5	10.0	26.1
12.50-25.00		22.7	19.8	19.2			13.8	36.1	11.6	29.8
25.00-50.00		25.9	17.3	22.6			16.7	43.6		33.2
>50.00		26.4		19.5			20.0	49.6		30.1
Rahimyar Khan	9.7	15.8	6.1*	22.5	54.0	7.9	11.0	41.8	9.3	
<12.50	9.2	15.6				8.1	10.2	37.2	6.0	
12.50-25.00	10.0	15.1	30.0		54.0	6.0	10.3	38.6	12.0	
25.00-50.00	9.2	13.7	12.5	22.5		9.0	10.4	44.0	12.0	
>50.00	12.5	16.8	3.5*			8.0	11.6	41.9	8.0	
Punjab	12.5	22.7	19.2	21.0	31.0	8.5	14.3	41.4	11.1	23.1
Jacobabad	8.7	8.4	17.7		23.6					
<12.50	7.7		18.1		30.0					
12.50-25.00	7.9	6.9	16.6		28.0					
25.00-50.00	5.8	6.8	18.0		23.9					
>50.00	10.0	9.2	17.8		22.9					
Larkana	7.2		23.6	19.1	37.5					
<12.50	5.0			17.0	39.1					
12.50-25.00	5.6				36.7					
25.00-50.00	6.6		20.0	18.6	36.0					
>50.00	8.5		25.4	20.0	38.5					
Nawabshah	9.6	32.9				13.4	13.2	673.4	400.0	
<12.50	9.6	35.5				10.9	13.7	679.0		
12.50-25.00	9.6	28.3				7.5	12.1	695.4		
25.00-50.00	9.0	29.9					11.9	677.1		
>50.00	10.0	36.2				20.0	14.0	666.4	400.0	
Hyderabad		20.2	14.8		21.9	16.4	19.9	609.7	27.5	
<12.50		18.6	18.8		26.4	12.6	16.2	608.6	40.0	
12.50-25.00		20.2	19.2		20.4	15.0	19.8	716.6	15.0	
25.00-50.00		17.4	17.0		22.6	11.5	20.3	609.2		
>50.00		21.5	13.0		21.2	18.4	20.2	584.9		
Sind	8.8	20.2	14.8	19.1	21.9	16.4	19.9	609.7	27.5	

*This figure appears unrealistic.

Notes: The output of sugarcane has been expressed in gur form in the Punjab and in cane form in Sind.

Blank spaces indicate data not applicable.

Source: Compiled by the author.

TABLE 4.4

Yield per Man-Day, Selected Crops
(in maunds, 82.28 lbs.)

District/	Wheat		Rice			Cotton		Sugar-	Maize	
Farm Size	Local	Mexi-Pak	Local	Basmati	IRRI	Local	Improved	cane	Local	Improved
Jhelum	0.1	0.2							0.1	
< 12.50	0.1									
12.50-25.00	0.1									
25.00-50.00	0.1								0.1	
> 50.00	200.0*	0.2								
Gujranwala	0.8	1.6	1.2	1.1	1.7	0.6		0.3	0.3	
< 12.50	0.6	1.1	1.3	1.4	1.9			0.3		
12.50-25.00	1.0	1.2	0.6	1.1	1.0	0.5		0.4	0.3	
25.00-50.00	0.7	2.0	1.1	1.4	2.2	0.7		0.3		
> 50.00	1.1	2.9	1.7	1.5	3.2			0.2		
Sahiwal	1.7	4.7	2.6	2.8	3.9	0.5	1.5	0.9	5.0*	1.9
< 12.50	1.4	1.5	1.5	1.5	4.3	0.3	0.6	0.6	5.0*	1.1
12.50-25.00	1.7	2.9	3.0	1.6	2.4	0.4	1.0	0.7		1.7
25.00-50.00	1.7	4.9	3.6	2.7	2.9	0.6	1.5	0.8		1.8
> 50.00	2.3	8.2	5.7	4.4	6.4	1.5	2.5	1.6		3.0
Lyallpur		2.5	0.4	0.7			0.8	1.3	0.4	0.9
< 12.50		1.5	1.2	0.7			0.3	0.6	0.4	0.7
12.50-25.00		1.4	0.5	0.3			0.4	0.5	0.6	0.6
25.00-50.00		2.6	0.6	0.6			0.8	1.3		1.1
> 50.00		4.9		1.1			1.5	2.6		1.8
Rahimyar Khan	0.4	0.5	0.3	0.9	1.9	0.4	0.5	1.5	0.5	
< 12.50	0.5	0.5				0.4	0.3	0.6	0.3	
12.50-25.00	0.4	0.6	1.0		1.9	0.3	0.4	1.2	1.2	
25.00-50.00	0.4	0.7	0.5	0.9		0.4	0.5	1.9	6.0*	
> 50.00	0.6	0.5	0.1			0.5	0.6	1.8	4.0*	
Punjab	0.4	2.0	1.1	1.5	1.8	0.3	0.8	1.0	0.2	1.2
Jacobabad	1.1	1.0	2.4		4.0					
< 12.50	0.5		1.0		0.5					
12.50-25.00	0.7	0.4	2.0		4.4					
25.00-50.00	1.1	0.6	4.5		3.8					
> 50.00	2.2	1.9	6.2		6.8					
Larkana	1.2		1.6	0.7	7.3*					
< 12.50	0.3			0.3	1.5					
12.50-25.00	0.5				3.9					
25.00-50.00	1.2		2.7	0.6	10.3*					
> 50.00	2.6		1.4	1.2	19.0*					
Nawabshah	1.5	5.4				0.8	1.7	101.7*	160.0*	
< 12.50	0.5	1.3				0.3	0.5	20.4		
12.50-25.00	0.7	4.3				0.4	0.6	40.6		
25.00-50.00	1.8	5.2					1.6	88.7*		
> 50.00	3.2	13.1				5.4	3.9	213.4*	160.0*	
Hyderabad		1.0	0.8		1.2	0.5	1.0	16.6	1.2	
< 12.50		0.9	0.7		1.1	0.5	0.5	12.7	2.7	
12.50-25.00		1.0	0.8		1.0	1.1	0.7	20.1	0.5	
25.00-50.00		1.1	0.9		1.4	0.4	0.8	17.5		
> 50.00		1.7	0.7		1.3	0.9	2.0	22.1		
Sind	1.3	1.0	0.8	0.7	1.2	0.5	1.0	16.6	1.2	

*This figure appears quite unrealistic.

Note: The output of sugarcane is expressed in gur form in the Punjab and in cane form in Sind. Blank spaces indicate data not applicable.

Source: Compiled by the author.

seeds in both provinces: 65 percent in the Punjab and 24 percent in Sind. The higher average labor productivity in the Punjab for both wheat and rice results from the greater use of farm machinery.

Looking at the interdistrict differences in the Punjab, the highest yields per acre and per man-day of both the new and the local seeds of wheat and rice are in Sahiwal. The only exception is the yield per acre of Mexi-Pak wheat, which is still higher in Lyallpur. The fact that the yield per acre of Mexi-Pak wheat is lower than that of local wheat in Jhelum is because it is a rain-fed district. The shortage of water in this area limits the use of the new wheat seeds.

In Sind, the highest yields per acre and per man-day of local and Mexi-Pak wheat are reported in Nawabshah. Also, the new wheat seeds in this district have significantly higher land and labor productivities. In rice the IRRI varieties have higher land and labor products than local varieties; Larkana and Jacobabad lead Hyderabad in both the new and the local seeds. Larkana leads the other two districts in the differentials of yields per acre and per man-day between the new and local varieties of rice. Larkana has the better position because its soil and water conditions are more suitable to the growth of rice.

Comparing the differences in yields per acre and per man-day of the Mexi-Pak varieties, it is evident that the yields increase with the size of farm in every district of the Punjab. The same pattern emerges for local wheat except in Sahiwal. There is, of course, no uniformity in the change between the districts. For IRRI rice the yield per acre and per man-day tends to increase with the size of farm only in Sahiwal. In Gujranwala there is no clear trend in the yield per acre, though the yield per man-day rises. There are no trends in Rahimyar Khan. For local rice it is only in Gujranwala that the yields increase with the size of farm. In Sahiwal, the yield per acre declines and the yield per man-day generally rises. For Basmati rice there seems to be an upward trend in the yield per acre and per man-day with the size of farm.

In Sind, the yields per acre and per man-day of Mexi-Pak wheat tend to increase with the size of farm in Jacobabad and Hyderabad. There are no clear trends in Nawabshah. For local wheat, there is an increase in land and labor productivity with the size of farm in all districts. For IRRI rice the yield per man-day rises with farm size, in all districts, while the yield per acre declines somewhat. In the case of local rice, in both Jacobabad and Hyderabad the yield per man-day increases, but the yield per acre has no specific trend in Jacobabad and falls in Hyderabad.

TABLE 4.5

Percentage Distribution of Crop Area Sown with New Seeds

District/ Farm Size	Mexi-Pak Wheat	IRRI Rice	Improved Cotton	Improved Maize
Jhelum	21.1			
< 12.50				
12.50-25.00				
25.00-50.00				
> 50.00	66.7			
Gujranwala	97.2	50.3		
< 12.50	98.8	48.7		
12.50-25.00	94.9	36.5		
25.00-50.00	98.0	46.3		
> 50.00	95.1	63.2		
Sahiwal	89.1	8.7	92.6	100.0
< 12.50	83.3	1.3	91.6	100.0
12.50-25.00	82.0	5.8	87.6	100.0
25.00-50.00	84.1	16.0	93.5	100.0
> 50.00	94.8	7.2	98.0	100.0
Lyallpur	100.0		100.0	97.8
< 12.50	100.0		100.0	87.0
12.50-25.00	100.0		100.0	94.8
25.00-50.00	100.0		100.0	100.0
> 50.00	100.0		100.0	100.0
Rahimyar Khan	79.5	1.9	83.7	
< 12.50	65.8		74.8	
12.50-25.00	70.7	33.3	82.1	
25.00-50.00	67.4		89.8	
> 50.00	96.1		82.6	
Punjab	91.7	31.0	93.7	98.8
Jacobabad	64.5	15.8	100.0	
< 12.50		2.0		
12.50-25.00	44.4			
25.00-50.00	72.3	18.2		
> 50.00	70.4	21.0	100.0	
Larkana		98.1		
< 12.50		100.0		
12.50-25.00		100.0		
25.00-50.00		97.7		
> 50.00		97.6		
Nawabshah	46.1		91.8	
< 12.50	29.6		73.4	
12.50-25.00	64.8		79.9	
25.00-50.00	48.5		100.0	
> 50.00	42.2		93.5	
Hyderabad	100.0	95.9	88.2	
< 12.50	100.0	93.1	90.7	
12.50-25.00	100.0	98.7	87.8	
25.00-50.00	100.0	92.8	82.9	
> 50.00	100.0	94.8	89.2	
Sind	89.6	61.4	90.8	

Note: Blank spaces indicate data not applicable.

Source: Compiled by the author.

DISTRIBUTION OF AREA SOWN WITH THE NEW SEEDS

As shown in Table 4.5, about 92 percent of the wheat area in the Punjab is planted with the Mexi-Pak varieties, as against 90 percent in Sind. But in rice there is a great difference: in the Punjab only about 31 percent of the rice area was used for IRRI varieties as against 61 percent in Sind. It is also clear that, in both provinces, over 90 percent of the cotton area was planted with the improved varieties. In maize, a crop that is not so dominant in Sind, 99 percent of the area in the Punjab was given to the improved varieties.

Taking the sample districts separately, in the Punjab only in Jhelum was a low percentage (21) of the wheat area under Mexi-Pak. In Rahimyar Khan, about 80 percent of the wheat grown was Mexi-Pak. For rice, since Gujranwala is a major district growing this crop, IRRI rice had accounted for only 50 percent.

In Sind, only Hyderabad had a high percentage of wheat area planted in Mexi-Pak. Even in Nawabshah the percentage was below 50. In rice, both Larkana and Hyderabad had more than 95 percent of the rice area devoted to IRRI varieties. Most of the area in Jacobabad grew local varieties.

In most districts where Mexi-Pak and IRRI varieties are grown, there is hardly any relationship between the percentage of the area given to these varieties and the size of holding. Only in one or two cases there is a discernible positive correlation of size and percentage of area under the new seeds: Jhelum (wheat), Rahimyar Khan (rice), and Jacobabad (rice).

CHAPTER

5

ADOPTION AND USE OF THE NEW INPUTS

To evaluate the adoption process and the present use of new inputs, it is necessary first to define the new inputs: chemical fertilizers, new seeds of wheat (Mexi-Pak) and rice (IRRI), pesticides, farm machinery, and tubewell water. It also is important to examine the time profile of adoption of the new inputs and the percentage of farmers using them.

ADOPTION OF NEW INPUTS

With the data in Tables 5.1 and A.5, the adoption process of individual inputs can be analyzed.

New Wheat and Rice Seeds

In the Punjab, Mexi-Pak wheat has been used at least once by 90 percent of the farmers, as against 64 percent in Sind. If the district of Jhelum in the Punjab is disregarded, because it is a rain-fed area and only a few farmers have used the new seeds, the percentage of farmers never using the new seeds would be even smaller. Also, it should be noted that a substantial percentage of the Punjabi farmers started using the new seeds earlier than did farmers in Sind: 1966 in the Punjab and 1968-69 in Sind. In the Punjab, during 1966-68 more than 72 percent of the farmers had adopted the new seeds. In Sind, on the other hand, it was during 1968-70 that about 51 percent of the farmers started the process of adoption.

TABLE 5.1

Adoption and Use of New Inputs

Input	% Never Used	% Used Once	Percentage Farmers Using 1972–73	2 Yrs.	4 Yrs.	6 Yrs.	8 Yrs.	Over 8 Yrs.	% Never Used	% Used Once	Percentage Farmers Using 1972–73	2 Yrs.	4 Yrs.	6 Yrs.	8 Yrs.	Over 8 Yrs.
	Jhelum								Gujranwala							
Fertilizer																
Nitrogenous	75.0	25.0	5.0		5.0	10.0		5.0	1.1	98.9			18.5	19.6	15.2	45.7
Phosphatic	100.0								67.4	32.6	1.1	1.1	5.4	4.4	3.3	17.4
Other	40.0	60.0		10.0			10.0	40.0	100.0							
New seeds																
Wheat (Mexi-Pak)	90.0	10.0	5.0	5.0					1.1	98.9			31.5	34.8	20.7	12.0
Rice (IRRI)	100.0									100.0		3.3	60.9	25.0	9.8	1.1
Pesticides	100.0								48.9	51.1		2.2	31.5	3.3	6.5	7.6
Farm machinery																
Tractor	70.0	30.0	15.0	5.0	5.0		5.0		75.0	25.0	1.1	2.2	17.4	2.2	2.2	
Other	70.0	30.0		5.0	15.0		10.0		100.0							
Tubewell water									38.1	62.0		2.2	20.7	18.5	13.0	7.6
	Sahiwal								Lyallpur							
Fertilizer																
Nitrogenous	0.6	99.5		1.6	16.3	29.3	26.6	25.5		100.0		1.4	18.9	42.6	28.4	8.8
Phosphatic	28.3	71.7	7.1	21.2	31.5	8.2	3.8		56.8	43.2		2.0	26.4	14.2	0.7	
Other	100.0								100.0							
New seeds																
Wheat (Mexi-Pak)	1.6	98.4		3.3	58.7	33.2	2.7	0.5	0.7	99.3		2.0	25.7	70.9		0.7
Rice (IRRI)	29.4	70.0		12.0	42.9	15.2		0.5	88.5	11.5			10.1	1.4		
Pesticides	73.9	26.1	4.4	15.8	3.3		2.7		51.4	48.6		30.4	18.2			
Farm machinery																
Tractor	65.8	34.2	2.2	12.5	17.4	1.6	0.5		67.6	32.4						
Other	100.0								99.3	0.7						
Tubewell water	33.7	66.3		2.2	23.4	15.2	8.7	16.9	71.6	28.4		5.4	16.9	4.7	0.7	0.7

(continued)

TABLE 5.1 (continued)

Input	% Never Used	% Used Once	Percentage Farmers Using 1972-73	2 Yrs.	4 Yrs.	6 Yrs.	8 Yrs.	Over 8 Yrs.	% Never Used	% Used Once	Percentage Farmers Using 1972-73	2 Yrs.	4 Yrs.	6 Yrs.	8 Yrs.	Over 8 Yrs.
	Rahimyar Khan								Punjab							
Fertilizer																
Nitrogenous		100.0	1.4	13.5	28.4	2.7	4.1	50.0	5.4	94.6	0.4	2.9	17.6	25.9	20.9	20.0
Phosphatic	71.6	28.4	6.8	9.5	8.1		2.7	1.4	30.3	69.7	3.7	9.7	20.9	7.7	2.5	3.3
Other									97.7	2.3		0.4			0.4	1.5
New seeds																
Wheat (Mexi-Pak)	20.3	79.7	6.8	37.8	16.2	4.1	14.9		9.5	90.5	1.2	7.3	35.3	37.5	6.8	2.5
Rice (IRRI)	100.0								55.7	44.2		4.8	27.0	10.2	1.7	0.4
Pesticides	56.8	43.2	5.4	10.8	10.8	1.4	6.8	8.1	63.3	36.7	2.3	16.2	11.8	0.8	3.1	2.5
Farm machinery																
Tractor	58.1	41.9	4.1	12.2	8.1	6.8	5.4	5.4	67.6	32.4	2.1	11.0	14.9	1.9	1.7	0.8
Other	100.0								98.7	1.4	0.6	0.2	0.4		0.2	
Tubewell water	37.8	62.2	9.5	17.6	14.9	8.1	4.1	8.1	49.0	51.0	1.4	5.2	18.3	11.2	6.2	8.7
	Jacobabad								Larkana							
Fertilizer																
Nitrogenous	8.5	91.5	6.4	25.5	55.3	4.3			2.1	97.9		17.0	57.5	21.3	2.1	
Phosphatic	93.6	6.4		4.3	2.1				78.7	21.3	2.1	8.5	8.5	2.1		
Other	100.0								100.0							
New seeds																
Wheat (Mexi-Pak)	6.4	93.6	14.9	53.2	25.5				57.5	42.6		27.7	14.9			
Rice (IRRI)	4.3	95.7	8.5	46.8	40.4				6.4	93.6		12.8	53.2	27.7		
Pesticides	97.9	2.1				2.1			57.5	42.6		17.0	10.6	10.6	4.3	
Farm machinery																
Tractor	57.5	42.6	4.3	10.6	10.6	12.8	4.3		51.1	48.9		27.7	19.2	2.1		
Other	97.9	2.1		2.1					95.8	4.3		2.1	2.1			
Tubewell water	95.8	4.3		2.1				2.1	100.0							

	Nawabshah								Hyderabad							
Fertilizer																
Nitrogenous		100.0		35.7	50.0	14.3			6.0	94.0		6.0	19.1	25.0	19.1	25.0
Phosphatic	82.1	17.9	1.8	12.5	3.6				95.2	4.8		1.2	2.4	1.2		
Other	100.0								98.8	1.2						1.2
New seeds																
Wheat (Mexi-Pak)	42.9	57.1		21.4	33.9	1.8			36.9	63.1	1.2	7.1	32.1	19.1	1.2	2.4
Rice (IRRI)	98.2	1.8			1.8				53.6	46.4	1.2	9.5	28.6	6.0	1.2	
Pesticides	7.1	92.9		26.8	61.5	8.9			86.9	13.1		2.4	4.8	2.4	1.2	2.4
Farm machinery																
Tractor	48.2	51.8	1.8	26.8	16.1	5.4			94.1	6.0		3.6	2.4			
Other	92.9	7.1		5.4	1.8				100.0							
Tubewell water	100.0								95.2	4.8		2.4	2.4			

	Sind							
Fertilizer								
Nitrogenous	4.3	95.7	1.3	19.2	41.5	17.5	7.3	9.0
Phosphatic	88.5	11.5	0.9	6.0	3.9	0.9		
Other	99.6	0.4						0.4
New seeds								
Wheat (Mexi-Pak)	36.3	63.7	3.4	23.9	27.8	7.3	0.4	0.9
Rice (IRRI)	44.9	55.1	2.1	15.4	29.5	7.7	0.4	
Pesticides	64.0	35.9		10.7	17.5	5.6	1.3	0.9
Farm machinery								
Tractor	67.1	32.9	1.3	15.4	10.7	4.3	0.9	0.4
Other	97.4	2.6	1.7	0.9				
Tubewell water	97.4	2.6	0.9	1.3				0.4

Note: Blank spaces indicate data not applicable.

Source: Compiled by the author.

FIGURE 5.1

Adoption of Mexi-Pak Wheat in the Punjab

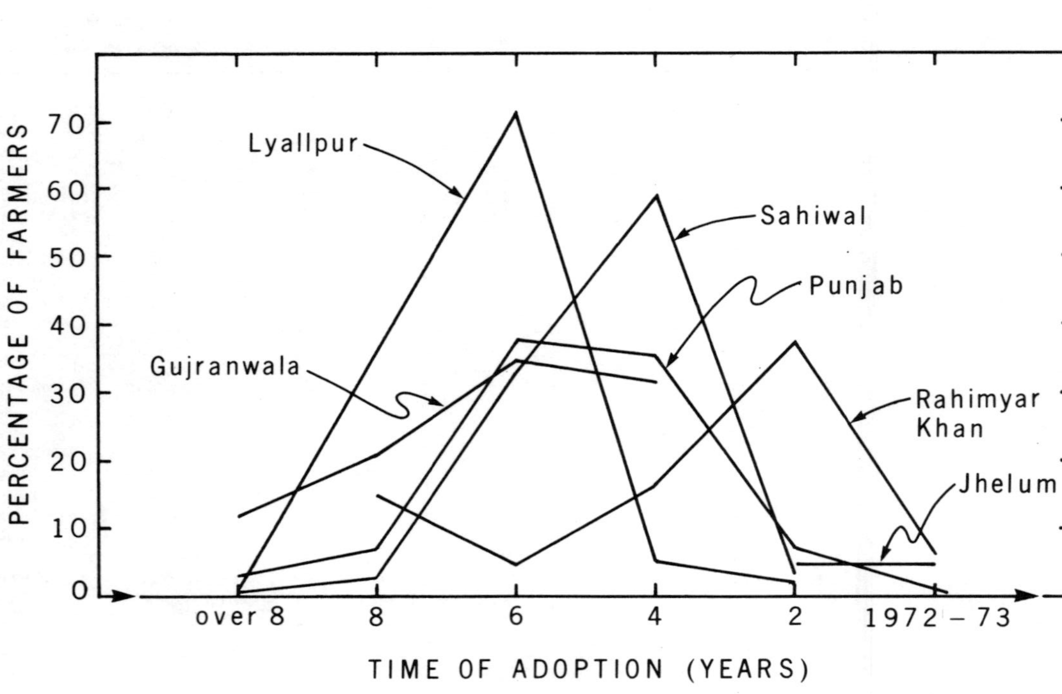

Source: Compiled by the author.

FIGURE 5.2

Adoption of Mexi-Pak Wheat in Sind

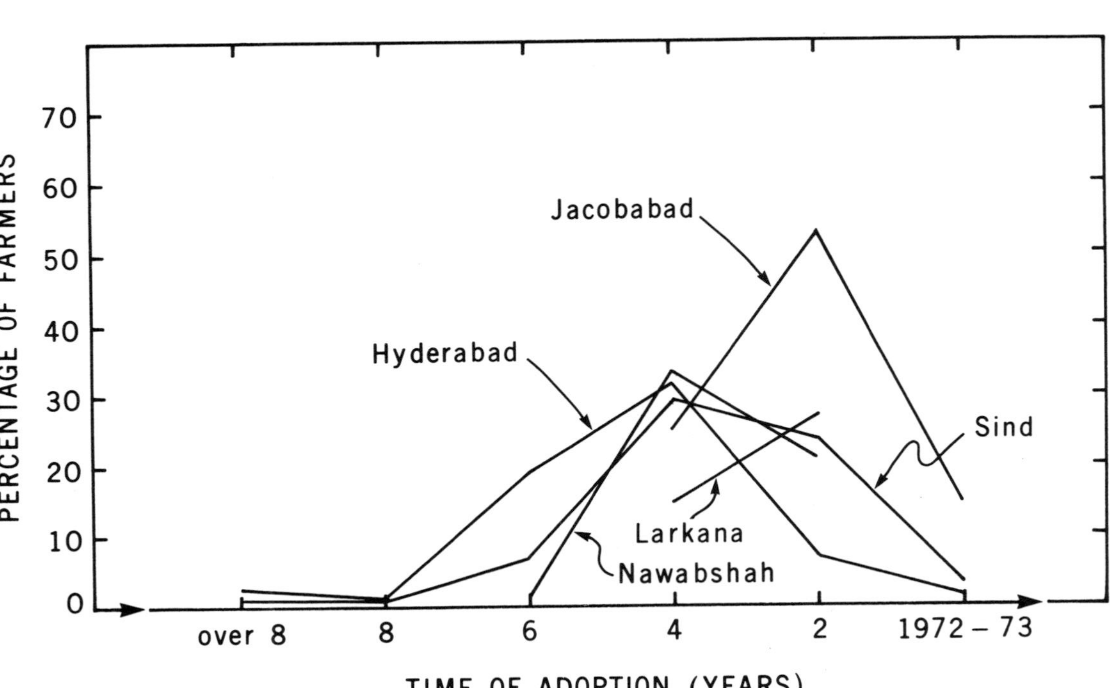

Source: Compiled by the author.

FIGURE 5.3

Adoption of IRRI Rice in the Punjab

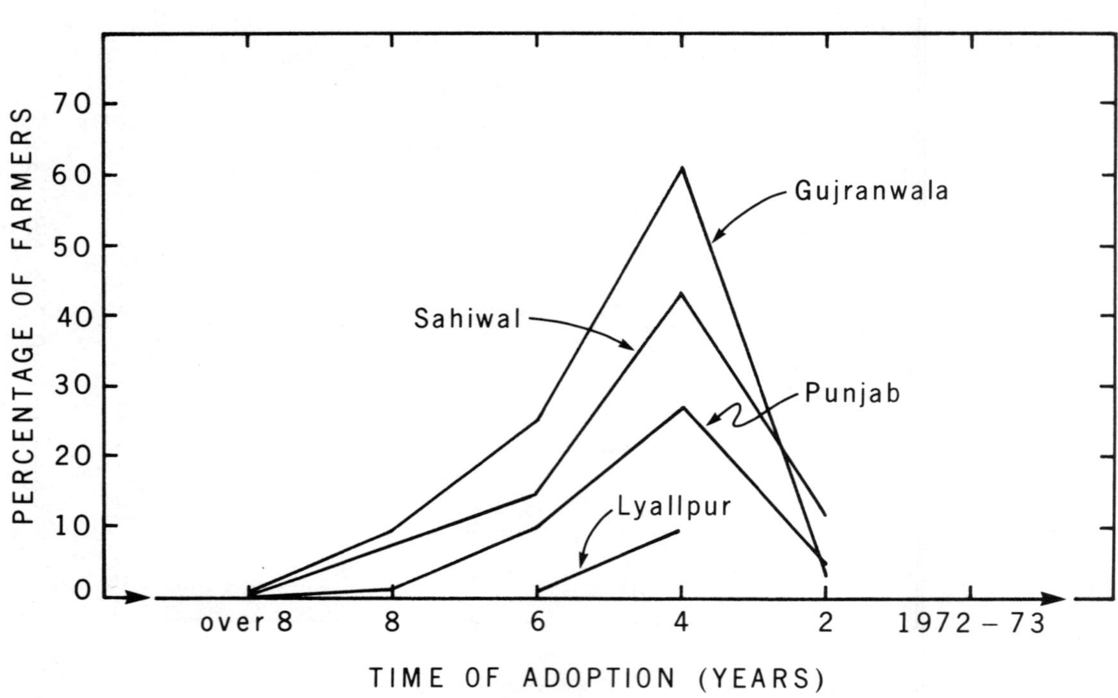

<u>Source</u>: Compiled by the author.

FIGURE 5.4

Adoption of IRRI Rice in Sind

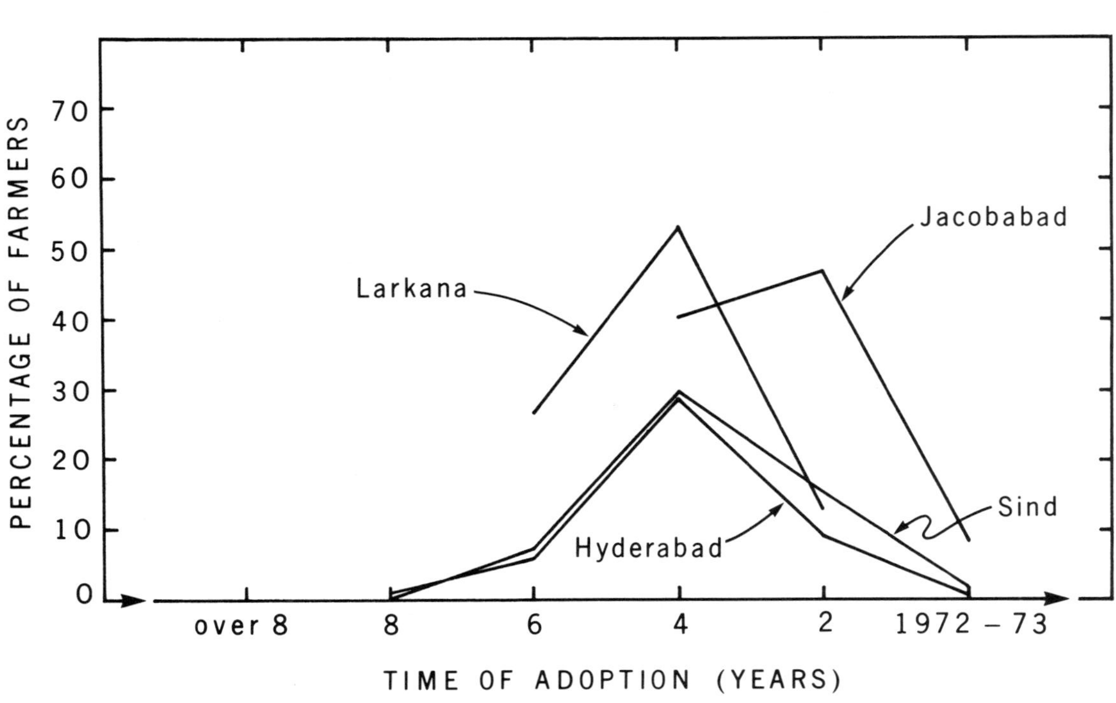

Source: Compiled by the author.

In the Punjab, the process of adoption started earlier in the relatively more progressive districts. Somewhat surprisingly, Gujranwala (a rice district) reported the earliest adoption. Jhelum and Rahimyar Khan reported adoption in 1968, but less than 80 percent of the farmers (in Jhelum a still smaller percentage) had used the new seeds. In the districts of Jhelum, Sahiwal, and Rahimyar Khan a higher percentage of farmers using the new seeds is associated with increased size of holding. There is no indication that in all districts the process of adoption began earlier on the large farms, though in Gujranwala, Lyallpur, and Rahimyar Khan larger farms started to use the new seeds somewhat earlier.

In Sind the process of adoption started later than in the Punjab and a smaller percentage of the farmers had used the new wheat seeds. The size of holding shows a more definite relationship with the percentage of farmers using the new seeds at least once. In Nawabshah and Hyderabad new seeds were adopted earlier by larger farms, while in Jacobabad and Larkana there seems to have been no difference. Hyderabad leads Sind in the adoption process.

About 55 percent of the farmers in Sind had used the IRRI varieties of rice at least once, but only 44 percent in the Punjab. In both provinces the IRRI varieties were first adopted in 1967-68.

Among the districts of the Punjab, Jhelum does not grow rice, and there is some IRRI rice in Rahimyar Khan. In Gujranwala, where all farmers used the IRRI varieties at least once, the adoption process began in 1968, with large farms apparently starting earlier than small ones. In Sahiwal, about 71 percent of the farmers had used the new seeds at least once; the adoption process began in 1968-70 and was not related to the size of the farm, although a higher percentage of large farms report growing the new seeds. In Lyallpur, where only 11 percent of the farmers reported ever using the new seeds, the process of adoption started in 1968. Large farms show a higher percentage using the IRRI seeds.

In Sind, Larkana and Jacobabad are the major rice districts in the sample, although in Hyderabad about 46 percent of the farmers had used IRRI seeds at least once, beginning in 1968-70. In Jacobabad and Hyderabad the large farms apparently started using the new seeds earlier, but in Larkana there is no discernible difference. Only in Larkana does there seem to be a difference in the percentage of farmers using the new seeds according to the size of the farm.

Chemical Fertilizers

The use of nitrogenous (N) fertilizers is higher in both provinces and started much earlier: 95 and 96 percent of the farmers in the Punjab and Sind had used N fertilizer at least once, as against 70 and 12 percent for phosphatic (P) fertilizer. A very small percentage had used any other type. The N fertilizer was used by almost an equal percentage of farmers in each province but the P fertilizer by a higher percentage in the Punjab than in Sind. Also, the use of N fertilizers had started earlier in the Punjab: more than 20 percent of the farmers had used them in the early 1960s. Both types of fertilizers have been used in Sind (except for N in Hyderabad) more recently, about two to six years. The earlier adoption of fertilizers in the Punjab can be explained by the longer tradition of commercial farming; the longer existence of, and more active participation by, public and private institutions (agricultural extension service, cooperative societies, and organized markets); the less feudal land-tenure system; and the development of the tubewell system for supplemental irrigation water. The same set of factors may explain the interregional differences within a province.

In the Punjab, more than 50 percent of the farmers in Rahimyar Khan had used N fertilizer at least once in 1964. In Sahiwal, over 25 percent had used it before 1964. In every district of the Punjab except Jhelum, almost all farmers had used N fertilizer at least once. In almost every district where it has been used, it was adopted on large farms earlier than on small farms. P fertilizer has been used at least once to varying degrees in the Punjab: Sahiwal and Rahimyar Khan are at the upper and lower ends, respectively. Its use in all districts, except Rahimyar Khan, started during 1966-68. As in the case of N fertilizer, earlier adoption is reported on the large farms, which also report a higher percentage of the farmers using P fertilizer at least once.

In Sind, although more than 25 percent of the farmers in most districts had used N fertilizer at least once, it was used before 1964 by a significant number of farmers only in Hyderabad. In Jacobabad, in fact, the adoption began in the last two to four years. As in the Punjab, the adoption process started earlier on the larger farms. For P fertilizer, again the size of the farm affects the percentage of farmers using it at least once and the time of adoption: the large farms report a higher percentage of farmers having used it at least once and also adopting it earlier. In Sind the use of P fertilizer started in 1969-70; only in Larkana were there farmers who had used it earlier.

Pesticides

The use of pesticides in both the Punjab and Sind started around 1968, and 36-37 percent of the farmers had used them at least once. It must, however, be noted that a higher percentage of farmers in Sind had used pesticides earlier.

In the Punjab, Jhelum is the only district where no respondent had ever used pesticides. Use appears to have started earliest in Rahimyar Khan. Although in Gujranwala more than 50 percent of the farmers had used pesticides at least once, there seems to be no difference in timing and use according to farm size. In all other districts there seems to be a positive relationship between the size of holding and the percentage of farmers having used pesticides.

In Sind Nawabshah leads other districts, with 93 percent of the farmers having used pesticides at least once. Also, in this district larger farms had begun the use earlier and a greater percentage had used pesticides at least once. In Jacobabad and Hyderabad, only the farmers with over 50 acres had ever used it. Larkana reports the earliest use of pesticides in Sind.

Farm Machinery

In both provinces the tractor is the dominant machine used on the farm and almost the same percentage of farmers (33 percent) have used it at least once. In almost every district the use of the tractor started around 1968 or later, and in most districts its use is directly related to the size of holding: on the larger farms use of the tractor is significant. In fact, except for Lyallpur (where use is reported by farmers holding over 12.50 acres), only the 50-acre or larger farms report the use of the tractor.

In the Punjab, rather surprisingly, Rahimyar Khan district leads the others: 42 percent of the farmers had used tractors at least once. In Sind, also rather surprisingly, Hyderabad had only 6 percent of the farmers reporting having used it. Nawabshah district reports the highest percentage, 52 percent.

Tubewell Water

The use of tubewell water is understandably greater in the Punjab: 51 percent of farmers in the Punjab report having used it at least once, as against only 3 percent in Sind. In Jhelum, no respondent had used tubewell water. The use of tubewell water in the Punjab started significantly during 1964-66; only in Rahimyar Khan is its use more recent. The size of farm is closely related to the use of tubewell water: the larger farms use it to a greater extent.

Sahiwal and Gujranwala, with 66 and 62 percent of the farmers using the tubewell water at least once, seem to have developed the tubewell system earlier than other districts. Although Rahimyar Khan is a late entrant, about 62 percent of the farmers there reported the use of tubewell water at least once.

USE OF SELECTED INPUTS

The survey has also attempted to quantify the level of use and the percentage of farmers who use such inputs as the new wheat and rice seeds, chemical fertilizers, irrigation water (canal, tubewell, surface well), and farm machinery. Tables 5.2 and A.6 give the data on the quantities of inputs used for wheat (local and Mexi-Pak), rice (local, Basmati, and IRRI), cotton (local and improved), and sugarcane.

Chemical Fertilizers

The use of chemical fertilizers in the Punjab and Sind differs in some interesting respects. A higher percentage of farmers in the Punjab seems to have used fertilizer on crops, except for cotton, but a somewhat greater amount of fertilizer per acre was used in Sind on the improved wheat, rice, and cotton. In general the use of fertilizer in the Punjab is much more diffused, and since Basmati rice is the most important variety, more fertilizer is used on it than on other varieties. Otherwise, in both provinces more farmers use a greater amount of fertilizer per acre on the high-yield varieties of wheat and rice and on improved cotton. In sugarcane, the difference in quantity used in the Punjab and in Sind is considerable. The explanation for the higher amounts of fertilizer used on a crop in an area can be found, for instance, in the higher cash return that the farmers expect. The interfarm differences may be explained by the financial means to buy fertilizer and the access to water.

In the Punjab, Rahimyar Khan leads in the amount of fertilizer used on Mexi-Pak wheat and Jhelum uses least. Also, Rahimyar Khan and Sahiwal lead in the amount of fertilizer per acre. In every district, farmers use more fertilizer on Mexi-Pak wheat than on local seeds. The same is true for the IRRI varieties of rice, except for Basmati in Gujranwala. The amount of fertilizer used increases with the size of farm in all districts, except for Mexi-Pak in Sahiwal. Only in Rahimyar Khan does the percentage of farmers using fertilizer tend to increase with the size of holding. In Gujranwala, Sahiwal, and Lyallpur, regardless of the size of holding, the percentage of farmers using fertilizer on Mexi-Pak and IRRI varieties is high.

In Sind, Hyderabad leads in the use of fertilizer: 2.86 bags per acre on Mexi-Pak wheat and 1.75 bags on IRRI rice. Jacobabad uses the smallest amount on IRRI and no fertilizer on Mexi-Pak wheat. Hyderabad also leads in the percentage of farmers using fertilizer on Mexi-Pak and IRRI varieties. Except for Hyderabad, where there seems to be some correlation between the size of holding and the amount of fertilizer used and the percentage of farmers using

TABLE 5.2

Use of Selected Inputs

Crop	Fertilizer		Irrigation Water: Canal		Irrigation Water: Sur. Well		Irrigation Water: Tubewell		New Seeds		Farm Machinery
	Av. No. Bags/Acre	% Farmers Using	Av. No. Irr./Ac.	% Farmers Using	Av. No. Irr./Ac.	% Farmers Using	Av. No. Irr./Ac.	% Farmers Using	% Farmers Growing Any Crop	% Farmers Growing This Crop	% Farmers Using
					Jhelum						
Wheat											
Local	0.1	5.6			2	5.6					27.8
Mexi-Pak	0.1	50.0							10.0	10.0	100.0
Rice											
Local											
Basmati											
IRRI											
Cotton											
Local											
Improved											
Sugarcane											
					Gujranwala						
Wheat											
Local	1.0	77.8	4	100.0			2	77.8			22.2
Mexi-Pak	1.9	86.0	5	87.1	3	1.1	3	47.3	100.0	91.1	21.5
Rice											
Local	0.9	100.0	12	89.5	4	5.3	6	94.7			21.1
Basmati	1.8	96.5	12	86.0	6	1.2	7	69.8			19.8
IRRI	1.1	98.8	12	84.9			8	67.4	93.5	45.0	23.3
Cotton											
Local	1.6	66.7	5	100.0			4	33.3			
Improved											
Sugarcane	2.0	68.8	17	96.9			5	71.9			18.8

	Sahiwal										
Wheat											
Local	1.4	100.0	4	100.0	1	1.3	2	76.0			30.7
Mexi-Pak	1.7	99.4	4	100.0	1	0.6	3	66.9	98.4	70.7	38.7
Rice											
Local	1.4	100.0	12	100.0	10	5.3	8	84.2			31.6
Basmati	2.0	99.4	14	100.0	8	1.3	9	73.5			36.8
IRRI	2.0	100.0	11	100.0	6	4.0	9	88.0	13.6	12.5	28.0
Cotton											
Local	1.0	26.7	9	100.0			5	60.0			33.3
Improved	1.6	60.4	8	100.0	3	1.2	6	66.9	91.9	92.4	38.5
Sugarcane	2.3	95.2	20	100.0	5	1.8	9	68.3			35.9
	Lyallpur										
Wheat											
Local											
Mexi-Pak	2.7	100.0	5	100.0			2	23.0	100.0	100.0	31.8
Rice											
Local	1.2	60.0	17	100.0							
Basmati	1.3	96.0	16	100.0			4	44.0			48.0
IRRI											
Cotton											
Local											
Improved	1.3	98.6	6	100.0			3	24.7	98.7	100.0	30.1
Sugarcane	2.3	99.3	16	100.0			5	29.7			32.4
	Rahimyar Khan										
Wheat											
Local	1.0	91.7	4	25.0	3	8.3	4	75.0			16.7
Mexi-Pak	3.1	89.7	4	50.0	4	3.4	5	56.9	78.4	71.6	36.2
Rice											
Local	1.0	14.3	10	85.7			6	28.6			28.6
Basmati	1.0	100.0	7	100.0							
IRRI	2.0	100.0	4	100.0			2	100.0	1.4	11.1	
Cotton											
Local	2.2	64.3	4	100.0			2	57.1			21.4
Improved	2.4	89.8	4	100.0			3	47.5	78.4	80.6	28.8
Sugarcane	3.2	88.7	10	100.0			5	46.8			22.6

(continued)

TABLE 5.2 (continued)

Crop	Fertilizer		Irrigation Water						New Seeds		Farm Machinery
			Canal		Sur. Well		Tubewell				
	Av. No. Bags/Acre	% Farmers Using	Av. No. Irr./Ac.	% Farmers Using	Av. No. Irr./Ac.	% Farmers Using	Av. No. Irr./Ac.	% Farmers Using	% Farmers Growing Any Crop	% Farmers Growing This Crop	% Farmers Using
					Punjab						
Wheat											
Local	1.3	83.3	4	71.4	2	3.2	3	65.1			27.0
Mexi-Pak	2.6	95.6	4	91.3	3	0.8	3	48.1	92.9	79.4	33.2
Rice											
Local	1.4	84.0	12	94.0	7	4.0	7	72.0			24.0
Basmati	2.1	98.1	13	95.5	7	1.1	8	69.3			32.2
IRRI	1.4	99.1	12	88.4	6	0.9	8	72.3	21.6	26.4	24.1
Cotton											
Local	2.1	46.9	7	100.0			4	56.3			25.0
Improved	2.3	79.9	6	100.0	3	0.8	5	47.3	72.0	92.3	33.7
Sugarcane	3.1	93.6	17	99.8	5	0.7	7	51.3			31.3
					Jacobabad						
Wheat											
Local			5	92.9			2	7.1			7.1
Mexi-Pak			7	90.5			3	9.5	44.7	60.0	4.8
Rice											
Local	0.4	59.1	20	95.5			18	6.8			6.8
Basmati											
IRRI	0.5	88.9	22	88.9			25	11.1	14.9	13.7	44.4
Cotton											
Local											
Improved									2.1	100.0	
Sugarcane											

	Larkana								
Wheat									
Local	0.5	15.4	3	84.6					
Mexi-Pak									
Rice									
Local	0.7	100.0	27	100.0					
Basmati	0.6	80.0	32	100.0					
IRRI	1.4	100.0	35	100.0			100.0	83.9	2.1
Cotton									
Local									
Improved									
Sugarcane									
	Nawabshah								
Wheat									
Local	0.5	11.5	2	100.0	2	3.8			
Mexi-Pak	1.6	100.0	9	94.4	3	5.6	32.1	40.9	5.6
Rice									
Local									
Basmati									
IRRI									
Cotton									
Local	1.4	100.0	11	100.0					
Improved	1.6	100.0	10	97.4	7	5.1	69.6	84.8	2.6
Sugarcane	1.8	100.0	22	96.9	13	6.3			3.1

(continued)

TABLE 5.2 (continued)

Crop	Fertilizer		Irrigation Water: Canal		Irrigation Water: Sur. Well		Irrigation Water: Tubewell		New Seeds		Farm Machinery
	Av. No. Bags/Acre	% Farmers Using	Av. No. Irr./Ac.	% Farmers Using	Av. No. Irr./Ac.	% Farmers Using	Av. No. Irr./Ac.	% Farmers Using	% Farmers Growing Any Crop	% Farmers Growing This Crop	% Farmers Using
					Hyderabad						
Wheat											
Local											
Mexi-Pak	2.9	76.5	6	100.0			3	8.8	81.0	100.0	8.8
Rice											
Local	1.0	100.0	20	100.0							11.1
Basmati											
IRRI	1.8	94.9	15	100.0					46.4	81.3	5.1
Cotton											
Local	3.5	50.0	8	100.0							10.0
Improved	2.0	84.6	7	100.0			3	12.8	46.4	79.6	12.8
Sugarcane	1.9	97.3	10	100.0			4	5.4			10.8
					Sind						
Wheat											
Local	0.5	11.4	3	91.1	3	1.3	2	2.5			1.3
Mexi-Pak	2.6	65.4	7	97.2			3	8.4	45.7	57.5	7.5
Rice											
Local	0.4	68.4	20	96.5			18	5.3			7.0
Basmati	0.6	80.0	32	100.0							
IRRI	1.6	96.8	26	98.9			25	1.1	39.7	60.0	7.4
Cotton											
Local	2.3	70.6	10	100.0							
Improved	1.9	92.3	8	98.7			4	9.0	33.8	82.3	7.7
Sugarcane	1.9	98.6	16	98.6			8	5.8			7.2

Note: Blank spaces indicate data not applicable.

Source: Compiled by the author.

it, there seems to be no definite relationship. In Sind, as in the Punjab, farmers use more fertilizer on the new varieties of both wheat and rice.

Irrigation Water

Sind leads the Punjab in both the amount of canal water and percentage of farmers using it on Mexi-Pak and IRRI varieties. While in the Punjab more or less the same amount is given to Mexi-Pak as to local wheat, in Sind the dosage to Mexi-Pak wheat is greater. For rice, however, the amount given to IRRI is less than to Basmati in both provinces.

In general, in the Punjab, except for Jhelum, where there is neither canal nor tubewell water, almost every district has a very high percentage of the farmers using canal water. But the amount of canal water is not necessarily affected by the size of holding, except in Rahimyar Khan, where there is some effect in the case of Mexi-Pak.

In Sind, where the dominant source of water is the canals, the districts differ quite significantly. In Jacobabad, where the amount of canal water is greater on the new varieties of wheat and rice, the percentage of farmers using this source is smallest. To some extent, it is made up by some tubewell and surface-well water. Larkana reports the greatest amount of canal water on IRRI rice, and all farmers use that source of water. In Nawabshah, which reports the highest amount of canal water on Mexi-Pak, the amount is significantly higher than on local wheat; it should be noted, however, that a smaller percentage of farmers use it on Mexi-Pak. This can be explained partly by the use of tubewell water on Mexi-Pak. In Hyderabad, almost all farmers use canal water on all crops they grow, but the dose on IRRI varieties of rice is lower than on local rice.

Tubewell water is used in the Punjab to supplement canal water, whereas in Sind, tubewells are an insignificant source of irrigation water. In the Punjab, the use of tubewell water is very significant in Rahimyar Khan, Sahiwal, and Gujranwala. In all districts except Rahimyar Khan, the use of tubewell water increases with the size of holding. However, it should be noted that in the tubewell-using districts, a lower percentage of farmers use tubewell water for the new varieties than local varieties of wheat and rice.

In Sind, where tubewell water is limited (Jacobabad, Nawabshah, and Hyderabad), a larger number of farmers use it for the new varieties of wheat and rice. As the size of farm increases, a higher percentage of farmers use tubewell water.

Surface wells are an important source of water only in Jhelum. In no other district is it a major source of irrigation water.

Considering irrigation water from all sources, the amount applied to Mexi-Pak wheat and IRRI rice is higher than on local varieties. This is, of course, quite consistent with the fact that more fertilizer is required for these varieties to realize their greater output potential.

New Seeds

In determining the extent to which the new seeds of wheat and rice are being used in the Punjab and Sind, it was decided to calculate the percentage of farmers using the new seeds among those who had grown any crop and those who had grown only the crop being considered. Eighty percent of the farmers growing wheat used the Mexi-Pak seeds in the Punjab, as against 58 percent in Sind. Of the farmers reporting any crop, 93 percent in the Punjab and 46 percent in Sind had used the new wheat seeds. From this it is obvious that Mexi-Pak wheat is grown more widely in the Punjab. For rice, however, the situation is reversed. In the Punjab, of the farmers reporting rice, only 26 percent had used IRRI seeds, as against 60 percent in Sind. Of the farmers reporting any crop, in the Punjab only 22 percent had used the IRRI seeds but 40 percent had in Sind. Most farmers in the Punjab and Sind who grow cotton used the improved seeds.

Analyzing the interdistrict and intercategory differences, Jhelum has very few farmers who have used the new wheat seeds. Only the farmers with 50 acres or more report their use. Lyallpur is the only district where all the responding farmers had used the new wheat seeds, followed by Sahiwal and Gujranwala; in Sahiwal, however, a higher percentage of farmers growing wheat reported the use of new seeds. In Rahimyar Khan more than 70 percent of the respondents had used the new wheat seeds. The use of new seeds in the Punjab does not show a uniform relationship to size of holding. Only in Jhelum and Rahimyar Khan is there a positive relationship between the size of holding and the percentage of farmers using the new wheat seeds. In fact, in Sahiwal the proportion of farmers using the new wheat seeds seems to decline as the size of farm increases. There is no effect of size on the use of new seeds in Lyallpur and Gujranwala.

Since the IRRI rice seeds are confined to Gujranwala in the Punjab--Sahiwal and Rahimyar Khan do not have a significant number of farmers growing rice--it should be noted that while most farmers grow IRRI rice, only 45 percent of those growing rice use the IRRI seeds. In Gujranwala the proportion of farmers growing IRRI rice declines with increased farm size.

In Sind, Hyderabad is the only district where all farmers growing wheat and more than 80 percent of all respondents use the new Mexi-Pak seeds. In Jacobabad 60 percent of the wheat farmers and 40 percent of all respondents use the new seeds. Although Nawabshah is an important wheat district, only 41 percent of the wheat farmers and 32 percent of the respondents had used the new seeds. In both Hyderabad and Nawabshah there is no correlation between size of holding on the use of new seeds, whereas in Jacobabad there is.

Larkana is the leading district for IRRI rice: all respondents and 84 percent of the rice farmers used the new seeds. In Hyderabad, although only 46 percent of the farmers grow IRRI rice, 81 percent of the rice growers used the IRRI seeds. In Jacobabad, where only 33 percent of the farmers grow IRRI rice, 26 percent of the rice farmers reported the use of IRRI seeds. The relationship between the size of the farm and the percentage of farmers using IRRI rice was positive in Jacobabad and negative in Larkana and Hyderabad.

Farm Machinery

The use of farm machinery is much more widespread in the Punjab than in Sind: 25-35 percent of the respondents in the Punjab had used farm machinery, as against only 7-8 percent in Sind. In both provinces the use of farm machinery was greater on Mexi-Pak than on local wheat; 33 percent of the farmers in the Punjab had done so, as against only 8 percent in Sind. In the Punjab, 24 percent of the farmers reported using farm machinery for IRRI rice, and in Sind only about 8 percent had used it. The use of farm machinery in the Punjab is more diffused; in Sind only those holding over 50 acres reported the use of farm machinery.

In the Punjab, a high percentage of farmers in Sahiwal, Lyallpur, and Gujranwala had used machines on Mexi-Pak wheat and IRRI rice. There is some relation between size of holding and use of machinery in every district. It is, however, on holdings of 50 acres or over that most farmers reported the use of farm machinery. A relatively small proportion of the farmers in Rahimyar Khan and Jhelum had used farm machinery on both wheat and rice.

In Sind, the use of farm machinery is reported by a higher percentage of the farmers in Hyderabad and Jacobabad. The relation between size of holding and use of machinery is more pronounced: the users are dominated by those who have at least fifty acres. In no district does any other farm size use machinery.

DISTRIBUTION OF OWNERSHIP OF TRACTORS AND TUBEWELLS

The use of tractors and tubewells by farmers in the Punjab and Sind is affected by their ownership. Punjabi farmers lead their Sindi counterparts in ownership, but in both provinces the ownership of tractors and tubewells is dominated by farmers who own fifty acres and more. In the Punjab, the farmers holding 25 acres and more also are well represented. Expenditures on the maintenance of tractors and tubewells in both provinces rise with the size of farm, as can be seen in Table 5.3.

Ownership of Tractors

In the Punjab, about 22 percent of the farmers own tractors, as against only 3 percent in Sind. The average cost of running a tractor in 1972-73 was higher in the Punjab than Sind: Rs. 6,037 versus Rs. 5,317. In both provinces there is a significant relationship between size of holding and tractor ownership. In fact, in Sind, only farmers holding 50 acres or more report the ownership of tractors. In the Punjab, except for Sahiwal and Rahimyar Khan, where there are some farmers in the 25-acre category who have a tractor, the owners are dominated by farmers with at least 50 acres.

Sahiwal leads the Punjab districts, followed by Lyallpur, Rahimyar Khan, Gujranwala, and Jhelum. It is the only district in the Punjab where only farmers having 50 acres or more own a tractor. In Sahiwal and Lyallpur, about 97 percent of the farmers owning at least 50 acres have a tractor. Although in Rahimyar Khan the distribution of ownership seems more diffused than in any other district of the Punjab, only 33 percent of the farmers with at least 50 acres own a tractor. Also in Rahimyar Khan the expenditure on tractors was the highest in the Punjab: Rs. 8,502. In every district the relationship between size of holding and tractor ownership is very strong.

In Sind, where the ownership of a tractor is not as common, Larkana reports no tractors. In Hyderabad, Jacobabad, and Nawabshah, only those farmers with holdings of over 50 acres or more own a tractor. In Hyderabad about 25 percent of the farmers in this category own a tractor. The maintenance expenditure on a tractor in Nawabshah was Rs. 1,500 in 1972-73, a figure that affected the average for Sind.

TABLE 5.3

Ownership of and Expenditure on Tractors and Tubewells, 1972-73

District and Farm Size	Tractors		Tubewells	
	Percent of Farmers Owning	Average Expenditure (in Rs.)	Percent of Farmers Owning	Average Expenditure (in Rs.)
Jhelum	10.0	2,000.0		
< 12.50				
12.50-25.00				
25.00-50.00				
> 50.00	10.0	2,000.0		
Gujranwala	8.7	3,362.5	28.3	1,473.1
< 12.50				
12.50-25.00			24.0	1,050.0
25.00-50.00	10.7	2,766.7	46.4	1,269.2
> 50.00	41.6	3,720.0	75.0	1,866.7
Sahiwal	30.4	3,107.1	53.3	1,897.3
< 12.50			2.2	1,000.0
12.50-25.00	4.1	2,000.0	42.9	2,107.1
25.00-50.00	23.3	2,900.0	79.1	1,617.9
> 50.00	95.7	3,204.0	93.5	1,992.0
Lyallpur	23.7	5,165.7	26.4	2,798.9
< 12.50				
12.50-25.00			2.6	1,100.0
25.00-50.00	5.3	4,000.0	10.5	1,387.5
> 50.00	97.1	5,236.0	100.0	3,161.8
Rahimyar Khan	14.9	8,501.8	21.6	5,065.0
< 12.50	5.0	5,000.0	5.0	5,500.0
12.50-25.00	5.0	6,500.0	10.0	2,225.0
25.00-50.00	20.0	7,375.0	45.0	4,968.9
> 50.00	33.0	10,504.0	26.7	5,065.0
Punjab	21.6	6,036.8	34.6	2,315.3
Jacobabad	2.1	6,400.0	2.1	9,500.0
< 12.50				
12.50-25.00				
25.00-50.00				
> 50.00	8.3	6,400.0	8.3	9,500.0
Larkana				
< 12.50				
12.50-25.00				
25.00-50.00				
> 50.00				
Nawabshah	1.8	15,000.0	3.6	8,200.0
< 12.50				
12.50-25.00				
25.00-50.00				
> 50.00	7.1	15,000.0	14.3	8,200.0
Hyderabad	6.0	3,164.0	3.6	3,866.7
< 12.50				
12.50-25.00				
25.00-50.00				
> 50.00	25.0	3,164.0	15.0	3,866.7
Sind	3.0	5,317.1	2.6	6,250.0

Note: Blank spaces indicate data not applicable.

Source: Compiled by the author.

Ownership of Tubewells

The percentage of farmers owning a tubewell is 35 and 3 in the Punjab and Sind, respectively. For 1972-73 the expenditure on a tubewell in the two provinces differed significantly: Rs. 2,315 in the Punjab and Rs. 6,250 in Sind. In both provinces the expenditure increases with the size of holding. While in both provinces there is a significant effect of the size of farm on the ownership of tractor, in the Punjab only farmers holding less than 12.50 acres do not own a tubewell. (Jhelum is the only district where there are no tubewells.) In Sind tubewells are owned only by farmers who have over 50 acres.

In the Punjab, Sahiwal leads other districts in that about 53 percent of the farmers owned a tubewell; 93 percent of the farmers having 50 acres or more owned a tubewell. In Lyallpur, although only 26 percent of the farmers owned a tubewell, all farmers owning at least 50 acres had one. In Gujranwala, where 28 percent of the farmers owned a tubewell, about 75 percent of the farmers holding a minimum of 50 acres were owners. In Rahimyar Khan, for some reason, a large percentage of the farmers having 25-50 acres owned a tubewell. Rahimyar Khan is also the most expensive district for running a tubewell (Rs. 5,065) and Gujranwala the least expensive (Rs. 1,473). In all districts the annual expenditure on tubewells increased with the size of holding.

In Sind, Larkana is the only district where no farmer owned a tubewell. In all others, tubewells were owned only by those who held at least 50 acres. The expenses of maintaining tubewells were Rs. 9,500 in Jacobabad, Rs. 8,200 in Nawabshah, and Rs. 3,867 in Hyderabad.

CHAPTER

6

THE DISTRIBUTION OF FARM OUTPUT BY USE

Farm output is used for seed, consumption, and market. The percentage distribution of output for these uses, given in Table 6.1, differs in the two provinces between farm sizes and crops.

A very high proportion of the Mexi-Pak wheat output is marketed in both provinces: 77 percent in the Punjab and 84 percent in Sind. Local wheat output in the Punjab is mainly consumed (93 percent); but in Sind only 23 percent is consumed and 70 percent is marketed. The percentage of marketed output of local and Mexi-Pak wheat is higher in Sind, which is explained by the fact that rice is the dominant staple crop.

In the Punjab most of the rice output is marketed: IRRI, 97 percent; local, 88 percent; Basmati, 90 percent. In Sind the marketed output is smaller for all rice varieties, especially Basmati: IRRI, 94 percent; local, 87 percent; Basmati, 51 percent. In both provinces IRRI rice dominates the marketed surplus. Since cotton and sugarcane are the two main cash crops in the Punjab and Sind, most of their output is marketed.

It is important to note that in both the Punjab and Sind, the marketed surplus of wheat and rice is affected by the size of the holding: as the size of holding increases, so does the marketed surplus. However, regardless of the size of holding, consumption requirements of wheat are met mainly by the local varieties. The consumption requirements of rice are met by local varieties in the Punjab and local and Basmati in Sind.

In each province the sample districts differ in the disposal of farm output. In the Punjab, in almost every district a very high percentage of the local wheat is consumed by the farmers. The marketed surplus of Mexi-Pak is highest in Sahiwal and lowest in Rahimyar Khan: 83 percent and 29 percent, respectively. Jhelum is exceptional

TABLE 6.1

Percentage Disposal of Farm Output, by Use

District/ Farm Size	Wheat: Local Seed	Wheat: Local Consumption	Wheat: Local Market	Wheat: Mexi-Pak Seed	Wheat: Mexi-Pak Consumption	Wheat: Mexi-Pak Market	Rice: Local Seed	Rice: Local Consumption	Rice: Local Market
Jhelum	11.6	86.2	2.2	4.8		95.2			
< 12.50	13.4	86.6							
12.50-25.00	14.4	85.6							
25.00-50.00	9.4	83.9	6.8						
> 50.0	10.0	90.0		4.8		95.2			
Gujranwala	4.6	84.8	10.6	5.0	17.6	77.4	0.4	11.9	87.7
< 12.50		100.0		6.3	54.3	39.4		37.2	62.8
12.50-25.00		100.0		5.4	25.7	68.9		62.5	37.5
25.00-50.00	6.7	89.5	3.8	5.3	19.3	75.4	0.5	15.8	83.7
> 50.00	5.1	77.2	17.7	4.4	7.5	88.2	0.3	3.4	96.3
Sahiwal	1.0	96.3	2.6	2.5	14.5	82.9	0.1	6.3	93.6
< 12.50	1.5	79.5	19.1	1.5	51.1	45.8		6.5	93.6
12.50-25.00	1.3	96.9	1.7	2.1	25.0	72.8	0.2	7.8	92.0
25.00-50.00	1.7	98.3		2.9	11.1	86.0	0.4	6.2	93.3
> 50.00		100.0		2.5	10.1	87.4		5.3	94.7
Lyallpur				4.1	16.8	79.1		36.7	63.3
< 12.50				5.1	52.0	42.9		66.7	33.3
12.50-25.00				4.7	31.5	63.8		38.1	61.9
25.00-50.00				4.2	20.1	75.7		32.7	67.3
> 50.00				3.9	10.1	86.0			
Rahimyar Khan	7.2	92.3	0.6	6.6	64.4	29.0	0.3	53.0	46.7
< 12.50	9.8	86.8	3.4	3.6	88.3	8.0			
12.50-25.00	4.0	96.0		6.4	80.6	13.1		33.3	66.7
25.00-50.00	7.3	92.7		5.7	80.8	4.5	0.8	70.6	28.6
> 50.00	9.6	90.4		7.2	50.5	42.3		45.9	54.0
Punjab	3.7	93.4	3.0	3.9	19.1	77.0	0.3	11.8	87.9
Jacobabad	12.9	25.3	61.9	10.1	20.4	69.5	6.0	6.5	87.5
< 12.50	10.6	29.2	60.2				6.6	16.8	76.6
12.50-25.00	16.7	27.8	55.6	7.2	54.3	38.5	7.6	12.0	80.5
25.00-50.00	12.2	33.5	54.3	7.0	32.5	60.5	5.7	7.1	87.1
> 50.00	12.8	22.5	64.8	11.2	13.7	75.1	5.6	3.5	90.9
Larkana	6.8	29.0	64.2				3.5	46.5	50.0
< 12.50	15.2	51.8	33.0						
12.50-25.00	10.8	48.3	40.9						
25.00-50.00	4.7	31.4	63.8					37.5	62.5
> 50.00	6.3	21.3	72.4				4.9	50.0	45.1
Nawabshah	4.9	16.2	78.9	2.1	5.2	92.7			
< 12.50	8.5	21.8	69.7	2.9	14.3	82.8			
12.50-25.00	11.1	29.3	59.6	3.5	7.4	89.2			
25.00-50.00	5.3	16.4	78.3	1.4	4.6	94.0			
> 50.00	2.8	12.7	84.6	2.0	3.4	94.7			
Hyderabad				3.1	16.7	80.2	1.7	16.8	81.5
< 12.50				1.2	25.0	73.8		47.1	52.9
12.50-25.00				2.9	27.0	70.1	4.8	20.8	74.4
25.00-50.00				3.1	18.2	78.7		15.2	84.8
50.00				3.4	12.7	83.9	1.9	11.5	86.5
Sind	7.00	23.5	69.5	3.3	12.5	84.1	5.8	7.3	86.9

Rice						Cotton		
Basmati			IRRI			Local		
Seed	Consumption	Market	Seed	Consumption	Market	Seed	Consumption	Market
0.8	12.7	86.6	0.3	2.9	96.8		18.9	81.1
3.1	19.9	77.0	0.1	13.8	86.1			
0.7	7.7	91.6	0.6	1.9	97.5			100.0
0.5	17.8	81.7	0.4	3.1	96.5		20.0	80.0
0.4	5.7	93.9	0.3	1.2	98.6			
0.0	6.7	93.3	0.3	1.0	98.7		2.9	97.1
0.0	14.8	85.1		6.7	93.3		7.8	92.2
0.0	11.4	88.5	0.4	4.0	95.6		4.2	95.8
0.0	7.6	92.4	0.3	1.4	98.3		2.3	97.7
0.0	4.2	95.8	0.3		99.7			100.0
	18.5	81.5						
	38.5	61.5						
	25.3	74.7						
	16.7	83.3						
	17.0	83.0						
	22.2	77.8		100.0			0.4	99.6
								100.0
				100.0				100.0
	22.2	77.8						100.0
							0.7	99.4
0.3	9.2	90.5	0.3	2.7	97.0		2.5	97.5
			4.6	1.7	93.7			
			8.3	3.3	88.3			
				5.7	94.3			
			4.5	2.6	92.9			
			5.7	0.9	94.0			
3.7	45.1	51.2	1.2	4.2	94.6			
5.9	94.1		2.5	12.5	85.0			
			1.7	7.1	91.2			
4.5	35.7	59.8	0.8	3.8	95.4			
2.8	41.7	55.6	1.0	2.4	96.5			
								100.0
								100.0
								100.0
								100.0
			1.1	8.7	90.4			100.0
				11.7	88.3			100.0
			0.1	11.5	88.4			100.0
				9.0	91.0			100.0
			2.1	6.8	91.2			100.0
3.7	45.1	51.2	1.4	4.8	93.8			100.0

(continued)

TABLE 6.1 (continued)

District/ Farm Size	Cotton (Improved)			Sugarcane			Maize (Improved)		
	Seed	Consumption	Market	Seed	Consumption	Market	Seed	Consumption	Market
Jhelum									
< 12.50									
12.50-25.00									
25.00-50.00									
> 50.00									
Gujranwala				2.1	14.8	83.2			
< 12.50				9.0	9.5	81.6			
12.50-25.00				0.4	5.2	94.4			
25.00-50.00				1.2	17.8	81.0			
50.00					23.8	76.1			
Sahiwal		0.8	99.2	0.1	6.0	94.0	2.5	1.5	96.0
< 12.50		3.5	96.5	0.1	9.7	90.3	1.6	4.5	93.8
12.50-25.00		1.1	98.9	0.1	8.3	91.7	2.5	2.5	95.1
25.00-50.00		0.9	99.1		7.0	93.0	2.2	2.1	95.7
> 50.00		0.4	99.6		3.1	96.9		0.4	96.9
Lyallpur		1.2	98.8	5.2	0.8	94.0	0.4	10.4	89.2
< 12.50	0.1	8.7	91.2	12.1	15.8	72.1		25.6	74.4
12.50-25.00		4.2	95.8	5.5	4.9	89.6	0.1	13.3	86.6
25.00-50.00		1.5	98.5	5.9		94.1	0.1	10.6	89.3
> 50.00		0.1	99.9	4.6		95.4	0.8	7.0	92.3
Rahimyar Khan		0.5	99.5	8.2	0.4	91.4			
< 12.50		0.8	99.1	5.0	6.2	88.8			
12.50-25.00		0.8	99.2	9.7	0.7	89.6			
25.00-50.00		0.8	99.2	10.2	0.3	89.5			
> 50.00		0.3	99.7	7.2		92.8			
Punjab		1.0	99.0	4.5	2.1	93.4	1.5	5.8	92.7
Jacobabad									
< 12.50									
12.50-25.00									
25.00-50.00									
> 50.00									
Larkana									
< 12.50									
12.50-25.00									
25.00-50.00									
> 50.00									
Nawabshah		0.5	99.5		1.6	98.4			
< 12.50		2.3	97.8		6.5	93.5			
12.50-25.00		0.6	99.5		2.7	97.3			
25.00-50.00		0.4	99.6		1.7	98.4			
> 50.00		0.3	99.7		1.2	98.8			
Hyderabad			100.0	0.9	0.1	99.0			
< 12.50			100.0		2.4	97.6			
12.50-25.00			100.0			100.0			
25.00-50.00			100.0			100.0			
> 50.00			100.0	1.8		98.2			
Sind		0.2	99.8	0.2	1.4	98.5			

Note: Blank spaces indicate data not applicable.

Source: Compiled by the author.

in that very little Mexi-Pak is grown and most of it is marketed. Where all varieties of rice are grown, IRRI leads in the marketed surplus. Again Sahiwal leads: more than 93 percent of the rice output is marketed. In Rahimyar Khan, where there is also some rice, the marketed surplus is relatively small: local, 47 percent; and Basmati, 78 percent.

The most striking feature in the Punjab is that the marketed surplus, except for local wheat, is influenced greatly by the size of holding. For example, in Lyallpur only 43 percent of the Mexi-Pak output is marketed by small farms (under 12.50 acres), and 86 percent by those of 50 acres and more. This strong and direct relationship between the size of holding and the percentage of grain output marketed is consistent with the argument that most small farmers are concerned with safety. For their cash needs, they would rather depend on crops other than those they must keep for consumption. Consumption of local wheat is higher among all farmers, which reflects the traditional preference for the flour from these varieties.

As was stated earlier, the marketed surplus of all grains is proportionately smaller in Sind. However, the same pattern emerges as in the Punjab: a higher percentage of the new varieties of both wheat and rice is marketed. In every sample district of Sind, as in the Punjab, there is a strong and positive relationship between the size of holding and the percentage of grain output marketed. In fact, this relationship is stronger in Sind.

CHAPTER

7

AVERAGE VALUE OF FARM OUTPUT AND ITS DISTRIBUTION

The concept of average value of farm output (AVFO) used in this study needs explanation. The value of output for each crop and on every farm has been computed by multiplying the reported quantity of crop output by the price that the farmer said he had received. For each farm category these values were added, and then the average for the responding farmers in each category was calculated. The average value for each district and province also has been calculated separately. Likewise the average value of farm output per acre has been calculated for each crop, farm category, and district. Finally, the percentage share of each major crop in the total value of farm output has been computed for each farm size in the sample districts.

AVERAGE VALUE OF FARM OUTPUT

The AVFO for the crops reported in the Punjab and Sind, given in Table 7.1, is Rs. 35,068 and Rs. 38,324, respectively. The higher value in Sind is explained mainly by the fact that the farm size in the sample is, on average, larger. In both provinces the AVFO is positively related to the size of farm: as the size of farm increases, the value of farm output increases. The highest value is reported in Nawabshah for holdings over 50 acres (Rs. 170,533), with Lyallpur in the Punjab next (Rs. 127,158). The lowest value of farm output for every farm size is in Jhelum.

In the Punjab, the AVFO among the sample districts differs significantly: Lyallpur (Rs. 46,700), Sahiwal (Rs. 40,000), Rahimyar Khan (Rs. 26,500), Gujranwala (Rs. 20,300), and Jhelum (Rs. 2,900).

TABLE 7.1

Average Value of Farm Output, Variable Cost, and Net Farm Income (rupees)

District/ Farm Size	Average Value of Farm Output	Average Variable Cost	Average Net Farm Income
Jhelum	2,898.45	664.71	2,233.74
< 12.50	1,093.94	264.25	829.69
12.50-25.00	3,335.29	499.18	2,836.11
25.00-50.00	4,412.83	884.33	3,528.50
> 50.00	6,316.00	2,516.50	3,799.50
Gujranwala	20,336.50	5,156.81	15,179.69
< 12.50	3,971.54	722.71	3,248.33
12.50-25.00	14,234.50	2,906.68	11,327.83
25.00-50.00	28,772.96	7,082.03	21,690.94
> 50.00	51,040.33	15,511.16	35,529.17
Sahiwal	39,999.21	6,273.97	33,725.24
< 12.50	9,507.89	1,708.92	7,798.97
12.50-25.00	21,432.57	3,408.45	18,024.11
25.00-50.00	40,786.29	6,448.57	34,337.73
> 50.00	88,270.10	13,529.94	74,740.16
Lyallpur	46,708.22	9,166.01	37,542.22
< 12.50	7,421.58	1,301.00	6,120.58
12.50-25.00	16,765.12	2,943.99	13,821.13
25.00-50.00	43,956.33	8,273.63	35,682.70
> 50.00	127,158.30	25,907.69	101,250.61
Rahimyar Khan	26,535.80	7,536.54	18,999.25
< 12.50	5,757.40	1,254.15	4,503.25
12.50-25.00	12,742.05	2,551.47	10,190.58
25.00-50.00	22,655.93	5,242.88	17,413.05
> 50.00	81,467.25	26,909.60	54,557.65
Punjab	35,068.05	6,865.65	28,202.40
Jacobabad	23,800.09	3,364.02	20,436.07
< 12.50	5,352.73	657.36	4,695.37
12.50-25.00	10,961.25	1,538.50	9,422.75
25.00-50.00	20,368.33	2,347.18	18,021.15
> 50.00	56,980.75	8,687.46	48,293.29
Larkana	39,668.18	4,662.79	35,005.39
< 12.50	8,191.18	1,227.03	6,964.15
12.50-25.00	22,455.06	2,964.49	19,490.57
25.00-50.00	41,475.25	5,335.21	36,140.04
> 50.00	83,928.15	8,838.13	75,090.02
Nawabshah	74,475.01	13,815.22	60,659.79
< 12.50	13,816.86	1,504.57	12,312.29
12.50-25.00	38,630.86	9,328.54	29,302.32
25.00-50.00	74,919.73	15,221.41	59,698.32
> 50.00	170,532.61	29,206.38	141,326.23
Hyderabad	22,054.17	4,051.20	18,002.97
< 12.50	5,703.07	846.32	4,856.75
12.50-25.00	11,615.05	1,920.23	9,694.82
25.00-50.00	19,869.76	3,926.74	15,943.02
> 50.00	53,707.83	10,045.10	43,662.73
Sind	38,324.08	6,345.59	31,978.49

Source: Compiled by the author.

In the Punjab, for the under 12.50-acre farms, the lowest AVFO (Rs. 1,093) is in Jhelum. Sahiwal has the highest average value for this category (Rs. 9,500), followed by Lyallpur (Rs. 7,422), Rahimyar Khan (Rs. 5,757), and Gujranwala (Rs. 3,972). For the farms of 12.50-25.00 acres, the highest value is reported in Sahiwal (Rs. 21,400), followed by Lyallpur (Rs. 16,800), Gujranwala (Rs. 14,200), and Rahimyar Khan (Rs. 12,700). The lowest value (Rs. 3,330) is in Jhelum. The highest value for farms of 25-50 acres is reported in Lyallpur (Rs. 43,956); then come Sahiwal (Rs. 40,800), Gujranwala (Rs. 28,800), Rahimyar Khan (Rs. 22,655), and Jhelum (Rs. 4,400). For farms over 50 acres, the highest value is reported in Lyallpur (Rs. 127,150), followed by Sahiwal (Rs. 88,270), Rahimyar Khan (Rs. 81,470), Gujranwala (Rs. 51,040), and Jhelum (Rs. 6,320).

Jhelum's having the lowest AVFO can best be explained by the limited number of crops, of low cash value, grown by the farmers in this rain-fed district. The high farm value in Lyallpur results from a greater specialization by farmers in cotton, wheat, and sugarcane--especially by farmers reporting over 25 acres. Rahimyar Khan's superiority to Gujranwala is accounted for by the inclusion of one very large farm (700 acres), which affects the average value of farm output very significantly.

In Sind, the highest value is in Nawabshah (Rs. 74,475); then come Larkana (Rs. 39,670), Jacobabad (Rs. 23,800), and Hyderabad (Rs. 22,050). For farms under 12.50 acres, Nawabshah has the highest value of output (Rs. 13,820), followed by Larkana (Rs. 8,190), Hyderabad (Rs. 5,700), and Jacobabad (Rs. 5,352). In the category of 12.50-25.00 acres, the highest average value is in Nawabshah (Rs. 38,630), then Larkana (Rs. 22,455), Hyderabad (Rs. 11,615), and Jacobabad (Rs. 10,960). For farms of 25 to 50 acres, Nawabshah again has the highest value (Rs. 74,920), followed by Larkana (Rs. 41,475), Jacobabad (Rs. 20,370), and Hyderabad (Rs. 19,870). Finally, for farms over 50 acres, Nawabshah still has the highest value of output (Rs. 170,530); then come Larkana (Rs. 83,930), Jacobabad (Rs. 56,980), and Hyderabad (Rs. 53,710).

Nawabshah's having the highest AVFO in Sind is explained by the fact that most farmers, especially those with more than 12.50 acres, specialize in sugarcane, cotton, and wheat. In no other district is there as much emphasis on cash crops. The low value in Hyderabad can be explained by the growth of a greater variety of crops, regardless of their cash return to the farmer.

AVERAGE VALUE OF FARM OUTPUT PER ACRE

A still better index of performance is the AVFO per acre. It eliminates the size effect on the value of output and also reflects the differentials between farm sizes and districts. The figures are given in Table 7.2 and Figure 7.1.

While the average value of farm output is higher in Sind, the value per acre is higher in the Punjab (Rs. 1,190, as against Rs. 1,100). Nawabshah is the highest-value district (Rs. 2,190 per acre), followed by Lyallpur (Rs. 1,630), which has the highest value per acre in the Punjab. Then comes Sahiwal (Rs. 1,305), and Larkana (Rs. 1,215). Jhelum is still the lowest-value district (Rs. 210). Gujranwala (Rs. 820) and Rahimyar Khan (Rs. 695) surpass Jacobabad (Rs. 665) and Hyderabad (Rs. 610).

In the Punjab--except for Jhelum, where the average value of output declines significantly as the size of farm increases--in every district the average value of farm output increases with the farm size. This trend is marked in Lyallpur and Sahiwal, and there is also an upward trend in Gujranwala. In Rahimyar Khan, because of the inclusion of one very large size farm in the over-50-acre category, there is some decline as the size of farm increases. The spread between the lowest and highest value of farm output per acre in the Punjab varies. In Jhelum the downward trend has a spread of over Rs. 110. In Sahiwal and Gujranwala the upward spread is about Rs. 200. In Lyallpur it increases to Rs. 1,000. In both Sahiwal and Lyallpur, there is a significant increase in the average value of farm output per acre between the two largest acreage categories. In all districts except Jhelum, the jump in value per acre from under 12.50 acres to 12.50-25.00 acres is about Rs. 45 to 50 per acre.

In Sind, where except in Larkana the upward trend in the AVFO per acre as the size of farm increases is not so stable, there is a significant upward trend, especially between the two largest-acreage groups. The largest spread in Sind is in Nawabshah (about Rs. 500), with Hyderabad at the lower end (Rs. 50). In all districts, the average value increases quite significantly when one moves from 25-50 acres to over 50 acres. The average value of farm output of the former category is lower than of the 12.50-25.00 acres category.

PERCENTAGE DISTRIBUTION OF AVFO BY CROP

In both provinces a very high percentage of the value of output is from the major crops: about 84 percent in the Punjab and 82 percent in Sind (see Tables 7.3 and A.7). However, it should be noted that the Kharif crops have a greater share of it in both provinces: 64 percent in the Punjab and 71 percent in Sind. The higher

TABLE 7.2

Average Value of Farm Output, Variable Cost, and Net Farm Income per Acre
(rupees)

District/ Farm Size	Average Value of Farm Output	Average Variable Cost	Average Net Farm Income
Jhelum	208.67	47.86	160.82
< 12.50	217.48	52.53	164.95
12.50-25.00	189.08	28.30	160.78
25.00-50.00	140.84	28.23	112.61
> 50.00	105.27	41.94	63.33
Gujranwala	818.70	207.60	611.10
< 12.50	706.68	128.60	578.08
12.50-25.00	753.55	153.87	599.68
25.00-50.00	792.21	194.99	597.22
> 50.00	931.56	283.10	648.46
Sahiwal	1,304.61	204.63	1,099.98
< 12.50	1,168.05	209.94	958.11
12.50-25.00	1,212.25	192.79	1,019.46
25.00-50.00	1,274.17	201.45	1,072.72
> 50.00	1,361.98	208.76	1,153.22
Lyallpur	1,630.30	319.93	1,310.37
< 12.50	912.86	160.02	752.84
12.50-25.00	1,063.78	186.80	876.98
25.00-50.00	1,463.75	275.51	1,188.24
> 50.00	1,973.28	402.04	1,571.24
Rahimyar Khan	696.29	197.76	498.53
< 12.50	656.49	143.00	513.49
12.50-25.00	694.39	139.04	555.35
25.00-50.00	722.68	167.24	555.44
> 50.00	691.04	228.26	462.78
Punjab	1,189.96	232.97	956.99
Jacobabad	667.23	94.31	572.92
< 12.50	577.43	70.91	506.52
12.50-25.00	578.13	81.14	496.99
25.00-50.00	577.82	66.59	511.23
> 50.00	740.01	112.82	627.19
Larkana	1,213.47	142.64	1,070.83
< 12.50	1,026.46	153.76	872.70
12.50-25.00	1,165.29	153.83	1,011.45
25.00-50.00	1,134.75	145.97	988.78
> 50.00	1,292.59	136.12	1,156.47
Nawabshah	2,188.51	405.97	1,782.54
< 12.50	1,746.76	190.21	1,556.55
12.50-25.00	2,234.29	539.53	1,694.76
25.00-50.00	2,083.99	423.41	1,660.58
> 50.00	2,275.28	389.68	1,885.60
Hyderabad	611.09	112.25	498.84
< 12.50	587.95	87.25	500.70
12.50-25.00	603.07	99.70	503.37
25.00-50.00	558.14	110.30	447.84
> 50.00	638.77	119.47	519.30
Sind	1,100.32	182.19	918.13

Source: Compiled by the author.

FIGURE 7.1

Average Value of Farm Output per Acre, by District and Farm Size

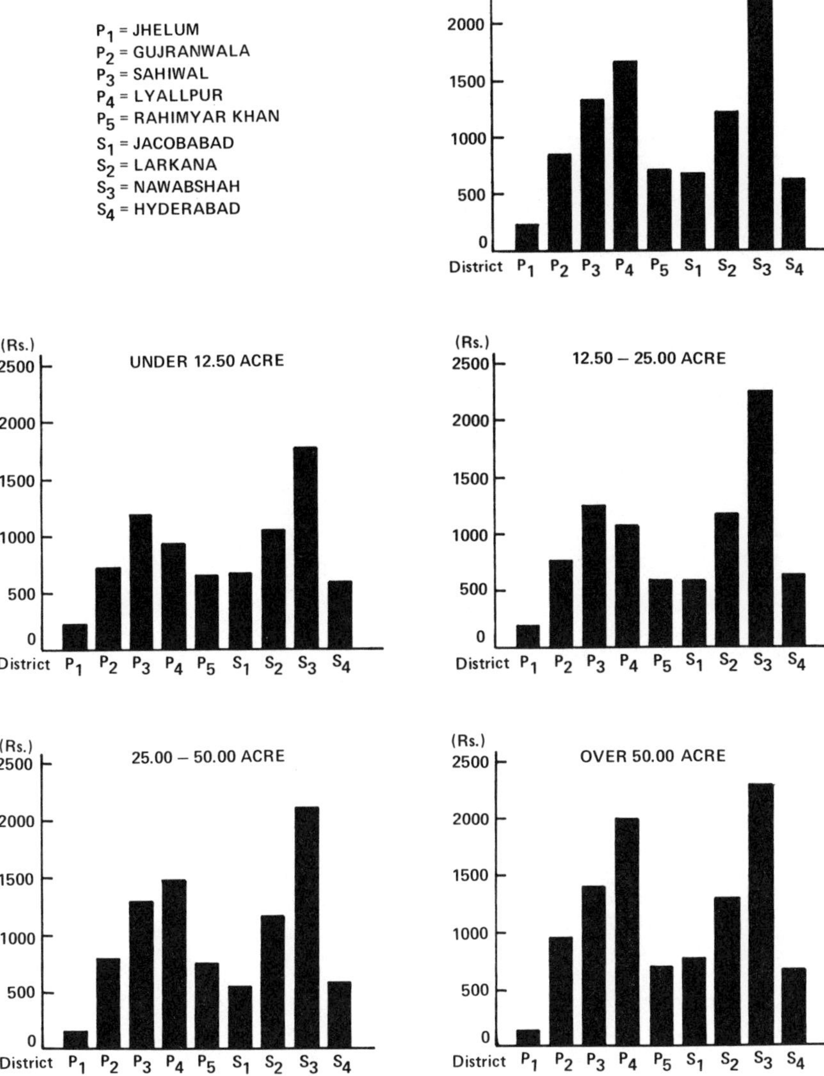

Source: Compiled by the author.

TABLE 7.3

Distribution of Average Value of Farm Output by Crop
(rupees)

Crop	Amount	(%)	Amount	(%)	Amount	(%)	Amount	(%)
	Jhelum		Gujrahwala		Sahiwal		Lyallpur	
Wheat								
Local	916.2	28.5	1,714.1	0.8	1,415.0	1.4		
Mexi-Pak	1,995.0	6.9	6,880.8	34.0	7,299.6	18.0	9,240.5	19.8
Rice								
Local			3,304.7	3.4	1,706.1	0.4	293.2	0.0
Basmati			5,155.6	23.7	6,481.5	13.7	1,475.1	0.5
IRRI			4,448.6	20.5	2,771.6	0.9		
Cotton								
Local			1,320.0	0.2	4,821.0	1.0		
Improved					14,539.0	33.4	16,201.7	34.2
Sugarcane			1,484.1	2.5	4,573.4	10.4	14,114.7	30.2
Maize								
Local	94.5	0.3	160.0	0.0	75.0	0.0	134.8	0.0
Improved					1,041.8	1.8	1,303.9	2.3
	Rahimyar Khan		Punjab		Jacobabad		Larkana	
Wheat								
Local	1,097.7	1.3	1,306.6	0.9	3,087.3	3.9	3,781.0	7.9
Mexi-Pak	3,782.1	11.2	7,369.5	19.6	3,206.2	6.0		
Rice								
Local	715.7	0.3	2,033.6	0.6	12,129.6	47.7	1,611.0	0.4
Basmati	2,250.0	0.1	5,569.8	8.2			2,617.0	0.7
IRRI	486.0	0.0	4,038.9	2.5	17,330.2	13.9	21,778.9	54.9
Cotton								
Local	7,604.3	5.4	5,710.5	1.0				
Improved	13,197.9	39.7	14,976.5	30.8				
Sugarcane	8,957.4	28.3	8,448.9	19.0				
Maize								
Local	112.0	0.0	121.9	0.0				
Improved			1,170.8	1.6				
	Nawabshah		Hyderabad		Sind			
Wheat								
Local	4,837.3	3.0			4,005.7	3.5		
Mexi-Pak	16,249.4	7.0	4,844.5	17.8	6,441.6	7.7		
Rice								
Local			1,655.0	0.8	9,737.6	6.2		
Basmati					2,617.0	0.2		
IRRI			6,648.0	14.0	15,145.8	16.0		
Cotton								
Local	11,138.0	1.9	4,913.9	2.7	7,476.8	1.4		
Improved	23,392.3	21.9	10,693.7	22.5	17,043.0	14.8		
Sugarcane	74,255.5	57.0	13,647.4	27.3	41,755.5	32.0		
Maize								
Local	12,000.0	0.3	590.0	0.0	4,393.3	0.2		
Improved								

Note: Blank spaces indicate data not applicable.

Source: Compiled by the author.

percentage of Kharif in Sind is, of course, contributed by sugarcane in Nawabshah and Hyderabad. In the Punjab, Kharif cotton is the major contributor. In Rabi crops, wheat contributes about 21 percent in the Punjab and only 11 percent in Sind. In both provinces, Rabi fodder, oilseeds, gram, and vegetables contribute a major part of the value of output.

In the Punjab, the crop contributions are wheat, 21 percent (Mexi-Pak about 20 percent); rice, 11 percent (of which Basmati contributes 8 percent and IRRI 3 percent); cotton, 32 percent (of which 31 percent is from improved seeds); and maize about 2 percent (most of which is from the improved varieties). In Sind, the crop contributions are wheat, 11 percent (of which about 8 percent is contributed by the Mexi-Pak varieties); rice, 22 percent (of which IRRI gives about 16 percent and local 6 percent); cotton, 16 percent (of which about 15 percent comes from the improved varieties); sugarcane, 32 percent.

The percentage contribution of each major crop to the AVFO differs in each district and also among the farm categories in the same district. In the Punjab, for example, the highest contribution to the farm value by these crops is made in Lyallpur (87 percent), followed by Gujranwala (85 percent), Rahimyar Khan (82 percent), Sahiwal (about 81 percent), and Jhelum (36 percent). It is obvious that with the exception of Jhelum, where most of the value comes from fodder and oilseeds, in no district do these crops together contribute less than 80 percent to the average farm value.

In Jhelum, 35 percent of the farm value is contributed by wheat (only 7 percent by Mexi-Pak). The percentage share of major crops increases with the size of farm from 34 percent to 62 percent, though there is some decline in the middle group.

In Gujranwala, the crop contributions are rice, 48 percent (Basmati 24 percent, IRRI 21 percent, and local 3 percent); wheat, 35 percent (Mexi-Pak, 34 percent); sugarcane, 3 percent. The percentage share of these crops in the AVFO moves from 85 to 91 percent, with some decline in the middle groups, as the farm size increases. Farm size also is related to the percentage contribution of individual crops. For example, both local and Mexi-Pak wheat increase their contribution with the size of farm. In rice there is very little change as the size increases. Sugarcane's contribution is reduced as the size increases.

In Sahiwal the share of cotton in the AVFO is about 34 percent (33 percent improved cotton); wheat, 19 percent (18 percent Mexi-Pak); rice, 15 percent (Basmati, 14 percent); sugarcane, 10 percent; maize, 2 percent. The collective share of these crops declines as the farm size increases, from 84 to 82 percent. In wheat, with the increased farm size Mexi-Pak increases and local declines.

There are no trends in rice, though there is some increase in the share of Basmati. Local cotton decreases and improved increases, but only for farms 50 acres and over. There is a decline in the share of sugarcane with increased farm size.

In Lyallpur, cotton and sugarcane dominate: cotton's share is 34 percent (no local cotton); sugarcane, 30 percent. Mexi-Pak wheat contributes about 20 percent. Maize (improved) makes a contribution of 2 percent. In this district there is a consistent increase in the share of these crops in the AVFO with the size of farm: from 75 to 88 percent. Wheat and maize decline as the size increases; sugarcane and cotton become the more important contributors.

In Rahimyar Khan, where 82 percent of the average farm value is contributed by the major crops, cotton and sugarcane dominate: cotton, 45 percent (40 percent from improved varieties); sugarcane, 28 percent. Wheat contributes about 12 percent (11 percent from Mexi-Pak). There is a consistent upward trend in the share of major crops in the AVFO as the farm size increases: from 80 to 87 percent. There is a definite increase in the shares of improved cotton and of sugarcane as the size increases. Local wheat declines but Mexi-Pak remains unchanged.

In Sind, wheat, rice, cotton, and sugarcane collectively contribute different percentages to the AVFO in the sample districts: Nawabshah, 91 percent; Hyderabad, 85 percent; Jacobabad, 72 percent; and Larkana, 64 percent. The share of major crops in the AVFO in Nawabshah exceeds that in all other districts of the Punjab and Sind; the inter-district spread is also larger in Sind.

In Jacobabad, rice dominates (62 percent, of which local rice contributes 48 percent and IRRI 14 percent); the other crop is wheat (10 percent, of which 6 percent is contributed by Mexi-Pak varieties). As the farm size increases, there is a consistent decline in the contribution of rice and wheat: from 78 to 69 percent. This decline results mainly from the local rice. Mexi-Pak wheat and IRRI rice increase significantly.

In Larkana, where the major crops contribute only 64 percent to AVFO, rice and local wheat dominate: 55 percent from IRRI rice and 8 percent from local wheat. There is a definite downward trend in the contribution by major crops, from 81 to 59 percent, as the size of farm increases. The decline is most noticeable in IRRI rice: from 71 to 50 percent.

Nawabshah is dominated by sugarcane (57 percent), followed by cotton (23 percent) and wheat (10 percent, of which 7 percent is from Mexi-Pak). In this district there is some increase in the contribution by major crops with the size of farm (from 85 to 90 percent), with a slight decline after 50 acres. Most of this increase is due to sugarcane, which increases its share from 18 to 57 percent. Cotton and wheat decline with increased size of farm.

In Hyderabad, where the major crops contribute about 85 percent to AVFO, sugarcane and cotton lead: cotton, 25 percent; sugarcane, 27 percent; Mexi-Pak wheat, 18 percent; and rice (IRRI), 14 percent. Although there is no discernible trend in the share of major crops as farm size increases, there is an overall decline from 89 to 82 percent. Cotton (improved) increases with farm size, whereas there is a significant decline in the share of IRRI rice from 22 to 11 percent. In sugarcane there is some increase followed by a decline. The share of Mexi-Pak wheat declines somewhat.

PERCENTAGE DISTRIBUTION OF AVFO PER ACRE, BY CROP

The AVFO per acre of the major crops, shown in Tables 7.4 and A.8, varies between the Punjab and Sind. For wheat and rice (except IRRI rice), it is higher in the Punjab. While the AVFO of local cotton is higher in Sind, that of the improved varieties is higher in the Punjab. The AVFO per acre of sugarcane is significantly higher in Sind. In both the Punjab and Sind, there is no clear trend in the AVFO per acre as the farm size increases.

In every district of the Punjab, except for Jhelum, the AVFO per acre of Mexi-Pak varieties is higher than of local wheat. Among rice varieties, the highest value per acre is derived from Basmati rice in all districts; but the contribution of IRRI varieties in each district is higher than of local rice. The AVFO of improved cotton varieties is significantly higher than that of local cotton. The highest value per acre of Mexi-Pak is reported in Sahiwal (Rs. 568), followed by Lyallpur (Rs. 567), Gujranwala (Rs. 426), Rahimyar Khan (Rs. 349), and Jhelum (Rs. 100). The highest value per acre of IRRI rice is in Rahimyar Khan (Rs. 972), followed by Sahiwal (Rs. 619), Gujranwala (Rs. 446), and Lyallpur (Rs. 358). In Basmati rice, Rahimyar Khan leads (Rs. 1,125), followed by Sahiwal (Rs. 973), Gujranwala (Rs. 628), and Lyallpur (Rs. 620). In the case of improved cotton, the highest AVFO per acre is reported in Sahiwal (Rs. 1,416), followed by Lyallpur (Rs. 1,277), and Rahimyar Khan (Rs. 972). Lyallpur leads in AVFO per acre of sugarcane (Rs. 2,178), followed by Rahimyar Khan (Rs. 1,802), Sahiwal (Rs. 1,504), and Gujranwala (1,484). In every district the effect of change in the size of farm is quite strong on the AVFO of local and Mexi-Pak wheat, Basmati and IRRI rice, and improved cotton: AVFO tends to increase with the size of farm.

In Sind, Nawabshah leads in the AVFO of all major crops it reports. The Mexi-Pak varieties of wheat have a higher value per acre than local varieties, and the differential is the largest in

TABLE 7.4

Distribution of Average Value of Farm Output per Acre, by Crop (rupees)

Crop	Jhelum	Gujranwala	Sahiwal	Lyallpur	Rahimyar Khan	Punjab	Jacobabad	Larkana	Nawabshah	Hyderabad	Sind
Wheat											
Local	110.3	301.2	384.5		209.1	272.8	240.1	216.7	296.4		246.4
Mexi-Pak	99.8	425.8	567.6	566.5	348.9	513.9	212.8		821.1	419.8	473.0
Rice											
Local		433.1	390.4		185.4	392.6	378.2	537.0		236.4	373.5
Basmati		628.0	973.2	619.8	1,125.0	826.4		769.7			769.7
IRRI		446.2	618.7	357.6	972.0	466.4	543.4	714.1		414.5	613.4
Cotton											
Local		416.4	803.5		680.2	713.8			1,218.6	773.8	996.9
Improved			1,415.7	1,276.7	971.9	1,275.7			1,212.7	880.2	1,084.2
Sugarcane		1,484.1	1,504.4	2,178.2	1,802.3	1,911.5			2,963.1	2,546.2	2,879.7
Maize											
Local	497.4	160.0	150.0	195.3	149.3	187.5			1,200.0	590.0	1,089.3
Improved			299.4	612.2		415.2					

Note: Blank spaces indicate data not applicable.

Source: Compiled by the author.

Nawabshah. IRRI rice leads local rice, although in Larkana (as in the Punjab), it has a lower value than Basmati rice. Of the two districts in which cotton is grown, the improved varieties have a higher value per acre in Hyderabad and a slightly lower value in Nawabshah. The highest value of Mexi-Pak wheat is in Nawabshah (Rs. 821), followed by Hyderabad (Rs. 420), and Jacobabad (Rs. 213). For IRRI rice, the highest value is in Larkana (Rs. 714), followed by Jacobabad (Rs. 543), and Hyderabad (Rs. 415). The highest value of sugarcane is in Nawabshah (Rs. 2,963), followed by Hyderabad (Rs. 2,546). In fact, Nawabshah has the highest value of Mexi-Pak and sugarcane among all the sample districts of the Punjab and Sind.

There is an interesting feature about the effect of change in the size of farm on AVFO per acre. For Mexi-Pak wheat, AVFO increases with the size of farm in every district. For IRRI rice, in every case it declines with increased farm size. In cotton, AVFO per acre increases with the size of farm in Nawabshah and Hyderabad. For sugarcane, however, in these two districts the value per acre tends to decline with increased farm size.

In both provinces, the value differentials of the local and new seeds of wheat, rice, and cotton need some explanation. In the case of wheat, the higher value of Mexi-Pak results from the higher yield per acre, because local and Mexi-Pak wheat have the same price per maund. The yield effect is also quite pronounced in the case of local and IRRI rice. The value per acre of Basmati rice is higher than that of IRRI and local varieties in both provinces because of the higher price per maund paid to the farmer for Basmati. In cotton, for which the price of local and improved varieties is the same in the two provinces, the higher yield of improved cotton is responsible for the higher AVFO per acre.

CHAPTER

8

AVERAGE VARIABLE COST AND ITS DISTRIBUTION

The concept of average variable cost (AVC), as defined in this study, includes the cost of seed; water from canal, tubewell (owned or leased), and surface well; chemical fertilizer, both nitrogenous and phosphatic; hired human labor (permanent and casual); farm machinery, which includes tractors and others (owned and leased); hired animal labor; and marketing costs, which include haulage and commission.

To determine the average cost of each of these factors, the total cost of each was calculated by multiplying the units used by the price per unit. The average was computed by dividing the total by the number of respondents. The AVC is tabulated for each farm category, district, and major crop. From the AVC for each crop, the cost per acre has also been determined. The percentage distribution of the AVC for each major crop is tabulated for each district and farm category. Finally, the average cost of each input for major crops and its percentage distribution has been calculated for the total area sown and on a per-acre basis.

AVERAGE VARIABLE COST

Before analyzing the AVC for different crops in the sample districts of the Punjab and Sind, it must be pointed out that the cost of canal water for each crop in Sind could not be determined. The reason is that the revenue officials in Sind do not levy a water tax by crop, as is done in the Punjab. There is a flat water rate: a certain cost per acre is levied regardless of the crops sown. The bias in favor of the AVC in Sind can, to some extent, be explained by this factor. The canal water costs for the Punjab and Sind are discussed in Appendix A.

The AVC, given in Table 7.1, is higher in the Punjab than Sind: Rs. 6,866 and Rs. 6,346, respectively. The highest AVC is in Nawabshah (Rs. 13,815), followed by Lyallpur (Rs. 9,166). The lowest is reported in Jhelum (Rs. 665). Nawabshah has also the highest AVC for each farm size. In almost all districts the AVC increases with the size of farm. This reflects the effect of size on the total variable cost.

In the Punjab, the district averages for the variable cost are Lyallpur, Rs. 9,166; Rahimyar Khan, Rs. 7,537; Sahiwal, Rs. 6,274; Gujranwala, Rs. 5,157; and Jhelum, Rs. 665. The dispersion of the AVC between the smallest and largest farm sizes is largest in Rahimyar Khan (Rs. 25,000), followed by Lyallpur (Rs. 24,000). The smallest dispersion is in Jhelum (Rs. 2,300). In Gujranwala and Sahiwal it is Rs. 11,000 and Rs. 12,500, respectively.

For farms under 12.50 acres, the highest AVC is in Sahiwal (Rs. 1,709), followed by Lyallpur (Rs. 1,301), Rahimyar Khan (Rs. 1,254), Gujranwala (Rs. 723), and Jhelum (Rs. 264). In the 12.50-25.00-acre size, the highest AVC is again in Sahiwal (Rs. 3,408), followed by Lyallpur (Rs. 2,944), Gujranwala (Rs. 2,907), Rahimyar Khan (Rs. 2,551), and Jhelum (Rs. 499). In the 25.00-50.00 acre category, Lyallpur reports the highest AVC (Rs. 8,274), followed by Gujranwala (Rs. 7,082), Sahiwal (Rs. 6,449), Rahimyar Khan (Rs. 5,243), and Jhelum (Rs. 884). At 50.00 acres and over, Rahimyar Khan takes the lead (Rs. 26,910), followed by Lyallpur (Rs. 25,908), Gujranwala (Rs. 15,511), Sahiwal (Rs. 13,530), and Jhelum (Rs. 2,517).

In Sind, the district AVCs are Nawabshah, Rs. 13,815; Larkana, Rs. 4,663; Hyderabad, Rs. 4,051; and Jacobabad, Rs. 3,364. The difference between the lowest and highest AVC within a district is largest in Nawabshah (Rs. 28,000), followed by Hyderabad (Rs. 8,500), Jacobabad (Rs. 8,000), and Larkana (Rs. 7,600).

For farms under 12.50 acres, Nawabshah leads the districts in Sind (Rs. 1,505), followed by Larkana (Rs. 1,227), Hyderabad (Rs. 846), and Jacobabad (Rs. 657). In the 12.50-25.00-acre size, the AVC is Nawabshah, Rs. 9,329; Larkana, Rs. 2,964; Hyderabad, Rs. 1,920; and Jacobabad, Rs. 1,539. Nawabshah leads in the 25.00-50.00-acre category (Rs. 15,221), followed by Larkana (Rs. 5,335), Hyderabad (Rs. 3,927), and Jacobabad (Rs. 2,347). Over 50.00 acres, Nawabshah again has the highest AVC (Rs. 29,206), followed by Hyderabad (Rs. 10,045), Larkana (Rs. 8,838), and Jacobabad (Rs. 8,687).

Nawabshah's having the highest AVC in Sind, as well as the highest net farm income, is explained mainly by the specialization of farmers in crops like sugarcane and cotton, which require higher levels of inputs than do other crops. The figures on AVC per acre, discussed in the next section, reinforce this observation.

AVERAGE VARIABLE COST PER ACRE

The AVC per acre (Table 7.2) is Rs. 233 in the Punjab and Rs. 182 in Sind. Nawabshah has the highest AVC per acre (Rs. 406), and the lowest AVC per acre is in Jhelum (Rs. 48). Rahimyar Khan, Sahiwal, and Gujranwala in the Punjab have higher AVC than Jacobabad, Larkana, and Hyderabad in Sind. AVC per acre does not show a definite trend with the change in size of farm.

In the Punjab, the AVC per acre is led by Lyallpur (Rs. 320), followed by Gujranwala (Rs. 208), Sahiwal (Rs. 205), Rahimyar Khan (Rs. 198), and Jhelum (Rs. 48). Under 12.50 acres the AVC is Sahiwal, Rs. 210; Lyallpur, Rs. 160; Rahimyar Khan, Rs. 143; Gujranwala, Rs. 129; and Jhelum, Rs. 53. On the 12.50-25.00-acre farms Sahiwal leads again (Rs. 193), followed by Lyallpur (Rs. 187), Gujranwala (Rs. 154), Rahimyar Khan (Rs. 139), and Jhelum (Rs. 28). In the 25.00-50.00-acre category, Lyallpur has the highest AVC (Rs. 276), followed by Sahiwal (Rs. 201), Gujranwala (Rs. 195), Rahimyar Khan (Rs. 167), and Jhelum (Rs. 28). Over 50.00 acres Lyallpur is again the leading district (Rs. 402), followed by Gujranwala (Rs. 283), Rahimyar Khan (Rs. 228), Sahiwal (Rs. 209), and Jhelum (Rs. 42).

The effect of farm size on the AVC per acre is seen clearly only in Gujranwala, where it increases steadily with farm size. In Lyallpur, the AVC first increases steadily, and then more sharply after 25 acres. In Sahiwal the AVC remains more or less stable with the change in size of farm. In Rahimyar Khan there is an overall increase in the AVC with the size of farm, although there is a decline from under 12.50 acres to 12.50-25.00 acres. In Jhelum the AVC declines significantly, although for the highest category it rises somewhat.

In Sind the AVC per acre is highest in Nawabshah (Rs. 406), followed by Larkana (Rs. 143), Hyderabad (Rs. 112), and Jacobabad (Rs. 94). Under 12.50 acres, Nawabshah leads (Rs. 190), followed by Larkana (Rs. 154), Hyderabad (Rs. 87), and Jacobabad (Rs. 71). In the 12.50-25.00-acre category, Nawabshah has the highest AVC (Rs. 540), followed by Larkana (Rs. 154), Hyderabad (Rs. 100), and Jacobabad (Rs. 81). At 25.00-50.00 acres, Nawabshah again reports the highest AVC (Rs. 423), followed by Larkana (Rs. 146), Hyderabad (Rs. 110), and Jacobabad (Rs. 67). Over 50 acres, the highest AVC is in Nawabshah (Rs. 390), followed by Larkana (Rs. 136), Hyderabad (Rs. 120), and Jacobabad (Rs. 113).

As in the Punjab, the size effect in Sind is not uniform; nor is it clear in every district. It is only in Hyderabad that the AVC per acre increases with the size of farm. Overall, there is the same relationship in Jacobabad. In Larkana, on the other hand, the AVC falls as the farm size increases. In Nawabshah, the AVC per acre first rises rather noticeably, then declines steadily.

FIGURE 8.1

Average Variable Cost per Acre, by District and Farm Size

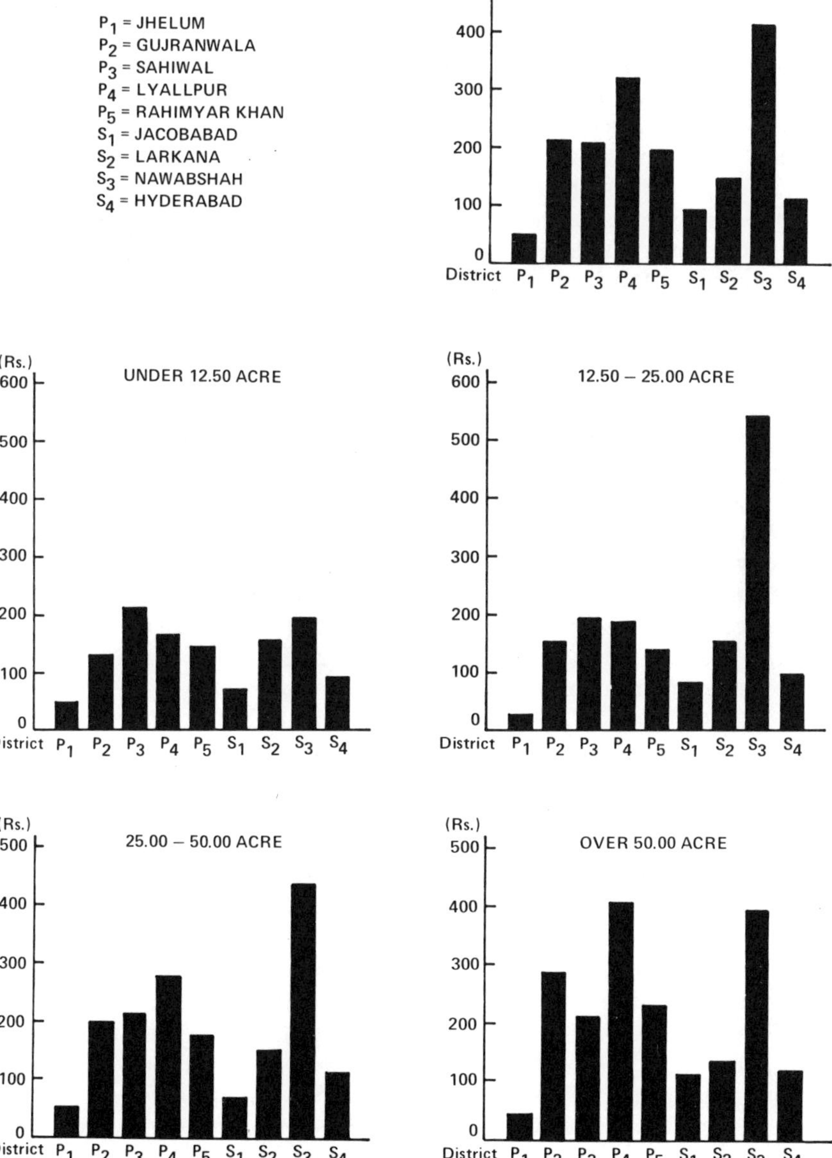

Source: Compiled by the author.

DISTRIBUTION OF AVERAGE VARIABLE COST BY CROP

The percentage share of the major crops (Tables 8.1 and A.9) differs significantly between the provinces: in the Punjab about 33 percent of the AVC is in wheat, as against only 14 percent in Sind. Mexi-Pak wheat in the Punjab claims 31 percent and 10 percent in Sind. Rice accounts for 21 percent of the total variable cost in Sind, as against only 11 percent in the Punjab. IRRI rice in Sind makes up about 16 percent of the total farm costs. Cotton in the Punjab has a share of over 21 percent, but only 10 percent in Sind. In both provinces the improved varieties of cotton dominate. Sugarcane claims a major portion of the farm costs in Sind: over 45 percent, as against 20 percent in the Punjab. The position of maize is rather insignificant. The major crops together in the Punjab make up over 87 percent of the farm costs and 90 percent in Sind.

The collective and individual shares differ widely among the sample districts in the Punjab and among the farm sizes in the same district. In Gujranwala the major crops claim over 90 percent of the total farm costs; in Jhelum, only 59 percent. For Sahiwal they are 83 percent; for Lyallpur, 89 percent; and for Rahimyar Khan, under 90 percent.

In Jhelum, of the major crops only wheat (local varieties) is grown; the rest of the total farm costs are accounted for mainly by fodder and oilseeds. There is no clear trend to the share of wheat in farm costs as the farm size increases, although for farms over 50 acres the percentage of wheat cost goes to 71 percent. Also, the share of local wheat in the total costs decreases significantly as the size of farm increases.

In Gujranwala, rice and wheat dominate in the total farm costs: 42 percent for wheat (40 percent for Mexi-Pak) and 45 percent for rice (23 percent for IRRI rice, 18 percent for Basmati, and 4 percent for local). Sugarcane accounts for just over 4 percent. The share of these major crops in the total farm costs changes very little with increased farm size: there is some increase in the share of Mexi-Pak wheat, some decline in sugarcane, and some addition in the share of IRRI rice.

In Sahiwal, where the share of major crops in total farm costs amounts to 83 percent, the share of wheat is 32 percent (29 percent for Mexi-Pak alone); cotton accounts for about 19 percent; sugarcane, 15 percent; rice, 14 percent; and maize, about 4 percent. In this district the collective share of major crops in total farm costs decreases with increased farm size, from 87 percent in the under-12.50-acres category to 83 percent in the over-50.00-acres size. The shares of individual crops in the costs change differently: while the share of local wheat declines, the share of Mexi-Pak wheat

TABLE 8.1

Distribution of Average Variable Cost, by Crop
(rupees)

Crop	Amount	(%)	Amount	(%)	Amount	(%)	Amount	(%)
	Jhelum		Gujranwala		Sahiwal		Lyallpur	
Wheat								
Local	263.6	35.7	810.9	1.5	443.1	2.9		
Mexi-Pak	1,552.5	23.4	2,062.4	40.4	1,858.1	29.1	2,989.7	32.6
Rice								
Local			966.6	3.9	534.9	0.9	126.7	0.1
Basmati			1,001.3	18.2	873.8	11.7	444.5	0.8
IRRI			1,266.6	23.0	670.7	1.5		
Cotton								
Local			378.6	0.2	616.8	0.8		
Improved					1,211.6	17.7	2,282.1	24.6
Sugarcane			529.1	3.6	1,025.9	14.8	2,586.8	28.2
Maize								
Local	15.6	0.2	127.6	0.0			32.8	0.0
Improved					347.8	3.8	290.4	2.6
	Rahimyar Khan		Punjab		Jacobabad		Larkana	
Wheat								
Local	622.7	2.7	477.9	1.7	454.7	4.0	666.8	11.9
Mexi-Pak	2,218.6	23.1	2,287.1	31.0	526.0	7.0		
Rice								
Local	302.4	0.4	625.6	0.9	1,588.9	44.2	172.9	0.3
Basmati	228.8	0.1	872.3	6.6			205.6	0.5
IRRI	80.2	0.0	1,123.0	3.5	2,756.7	15.7	3,273.5	70.2
Cotton								
Local	1,638.0	4.1	1,041.2	0.9				
Improved	3,172.7	33.6	1,938.9	20.4				
Sugarcane	2,343.5	26.1	1,751.6	20.1				
Maize								
Local	24.4	0.0	32.4	0.0				
Improved			319.6	2.3				
	Nawabshah		Hyderabad		Sind			
Wheat								
Local	600.7	2.0			607.4	3.2		
Mexi-Pak	2,710.7	6.3	1,386.3	27.7	1,440.3	10.3		
Rice								
Local			606.2	1.6	1,334.4	5.1		
Basmati					205.6	0.1		
IRRI			1,393.1	16.0	2,452.6	15.6		
Cotton								
Local	916.2	0.8	586.4	1.7	722.2	0.8		
Improved	2,255.7	11.4	1,327.2	15.2	1,791.5	9.4		
Sugarcane	17,908.9	74.1	2,833.8	30.8	9,825.2	45.5		
Maize								
Local	590.0	0.1			216.3	0.0		
Improved								

Note: Blank spaces indicate data not applicable.

Source: Compiled by the author.

increases. The share of both Basmati and IRRI rices increases with size, as does the share of improved cotton.

In Lyallpur, where major crops account for about 89 percent of total farm costs, the share of wheat (which is all Mexi-Pak) is 33 percent. Then come sugarcane with 28 percent, and improved cotton with 25 percent. Rice and maize are insignificant. In this district there is a steady decline in the share of major crops in the total farm costs as farm size increases: from over 93 percent for under 12.50 acres to 88 percent for over 50 acres. The only crop whose share in the total costs increases with farm size is Mexi-Pak wheat.

In Rahimyar Khan, where just under 90 percent of the farm costs are represented by the major crops, cotton accounts for 38 percent (improved cotton, 34 percent); then come sugarcane and wheat (mostly Mexi-Pak), with about 26 percent. The shares of rice and maize are negligible. In this district, too, there is some decline in the collective share of major crops in the total farm costs as the size of farm increases. For individual crops, the share of Mexi-Pak remains stable but that of improved cotton increases. There is some increase for sugarcane, but its share declines in the over-50.00-acres category.

In Sind, while the major crops collectively account for 90 percent of the total farm costs, their shares differ considerably among the districts and within each district among the various farm sizes. Nawabshah leads, with about 95 percent of the total farm costs. Jacobabad, with 71 percent, is lowest. Hyderabad and Larkana report 93 and 83 percent, respectively.

In Jacobabad only rice and wheat are reported among the major crops. The share of rice in the total farm costs amounts to about 60 percent (local rice accounting for 44 percent and IRRI rice for 16 percent); the share of wheat is 11 percent (Mexi-Pak, 7 percent, and local, 4 percent). The collective share of these crops increases with the farm size from about 68 percent to 73 percent. Individually, the shares of both IRRI rice and Mexi-Pak wheat in total farm costs increase with farm size. Both local rice and local wheat register a definite decline.

In Larkana, where rice (especially IRRI varieties) and wheat (local) are the two major crops, the share of rice in total farm costs is about 71 percent (IRRI rice, 70 percent) and of local wheat, about 12 percent. The share of wheat and rice together in the farm costs drops from 87 percent to 81 percent as farm size increases. The share of IRRI rice decreases with increased farm size, although there is initially some increase; the share of local wheat remains stable.

Nawabshah, which leads the other districts in Sind, has three major crops: sugarcane (74 percent of the total costs), cotton (12 percent; 11 percent improved cotton); wheat (8 percent; Mexi-Pak wheat, 6 percent). There is no other district in the sample in which one crop is as dominant as sugarcane is in Nawabshah. The collective share of these three crops increases with farm size from 87 percent to 94 percent, although there is some decline between 25.00-50.00 acres and over 50.00 acres. Sugarcane is the only crop in which the individual share in total farm costs increases significantly and then registers some decline. The shares of local and Mexi-Pak wheat and of local and improved cotton decrease as farm size increases.

In Hyderabad, where the major crops constitute about 93 percent of the total farm costs, sugarcane dominates with 31 percent. Wheat (Mexi-Pak) accounts for 28 percent of the costs; rice has a share of 18 percent (IRRI rice, 16 percent); and cotton's share is 17 percent (15 percent for improved cotton). With increased farm size, the collective share of major crops in total farm costs declines from 98 percent to 91 percent. Individually, only in the case of IRRI rice is there a steady decline in its share as the size of farm increases; other crops have no definite trend.

DISTRIBUTION OF AVERAGE VARIABLE COST PER ACRE BY CROP

The figures for the AVC per acre of crops are given in Tables 8.2 and A.10. Comparing the Punjab and Sind, the most striking feature is that, except for sugarcane, the AVC per acre for every crop is higher in the Punjab. In fact, in almost every case the difference is considerable. One reason, although by no means the most important one, may be that the AVC cost of crops in Sind does not include the cost of canal water. The more important reason is perhaps the greater use of modern inputs in the Punjab. With few exceptions, such as sugarcane in Hyderabad, in both provinces the AVC per acre increases with farm size. More interesting are the cases, by no means limited, in which there is a considerable increase in the AVC per acre as one moves from under 50 acres to over 50 acres.

In the Punjab, the highest AVC per acre for local wheat is registered in Gujranwala (Rs. 140), followed by Sahiwal (Rs. 120), Rahimyar Khan (Rs. 119), and Jhelum (Rs. 32). In the case of Mexi-Pak wheat, the highest cost per acre is reported in Rahimyar Khan (Rs. 205), followed by Lyallpur (Rs. 183), Sahiwal (Rs. 144), Gujranwala (Rs. 128), and Jhelum (Rs. 78). Only in Gujranwala is

TABLE 8.2

Distribution of Average Variable Cost per Acre, by Crop
(rupees)

Crop	Jhelum	Gujranwala	Sahiwal	Lyallpur	Rahimyar Khan	Punjab	Jacobabad	Larkana	Nawabshah	Hyderabad	Sind
Wheat											
Local	31.7	140.3	120.4		118.6	99.8	35.4	38.2	36.8		37.4
Mexi-Pak	77.6	127.6	144.5	183.3	204.7	159.5	35.0		137.0	120.1	105.8
Rice											
Local		126.7	122.4	154.5	78.4	120.8	49.6	57.6		86.6	51.2
Basmati		122.0	131.2	186.8	114.4	129.4		60.5			60.5
IRRI		127.0	149.7		160.4	129.7	86.4	107.3		86.9	99.3
Cotton											
Local		119.4	102.8		146.5	130.2			100.2	92.3	96.3
Improved			118.0	179.8	233.6	165.2			116.9	109.2	114.0
Sugarcane		529.1	337.5	399.2	471.5	396.3			714.6	528.7	677.6
Maize											
Local	82.1	127.6		47.5	32.5	49.9			59.0	29.5	54.1
Improved			100.0	136.4		113.3					

Note: Blank spaces indicate data not applicable.

Source: Compiled by the author.

the cost per acre of local wheat higher than that of Mexi-Pak; in other districts, where both varieties are grown, the cost of Mexi-Pak is higher. The AVC per acre of local rice is highest in Lyallpur (Rs. 155), followed by Gujranwala (Rs. 127), Sahiwal (Rs. 122), and Rahimyar Khan (Rs. 78). For Basmati rice, the highest AVC is reported in Lyallpur (Rs. 187), followed by Sahiwal (Rs. 131), Gujranwala (Rs. 122), and Rahimyar Khan (Rs. 114). In the case of IRRI rice, the highest AVC per acre is in Rahimyar Khan (Rs. 160), followed by Sahiwal (Rs. 150) and Gujranwala (Rs. 127). Gujranwala appears to be the lowest-cost district for rice. In almost every district the IRRI rice costs more to produce per acre than other varieties. The cost differential between the rice varieties in Gujranwala is the smallest. The AVC per acre of local cotton is highest in Rahimyar Khan (Rs. 147), followed by Gujranwala (Rs. 119), and Sahiwal (Rs. 103). For improved cotton, the highest AVC is in Rahimyar Khan (Rs. 234), followed by Lyallpur (Rs. 180) and Sahiwal (Rs. 118). In both Sahiwal and Rahimyar Khan, the AVC of improved cotton is higher than that of local varieties. In sugarcane, the highest AVC per acre is in Gujranwala (Rs. 529), followed by Rahimyar Khan (Rs. 472), Lyallpur (Rs. 399), and Sahiwal (Rs. 337).

In Sind, where the AVC per acre of every crop (except sugarcane) is lower than the Punjab, Nawabshah reports the highest AVC per acre of most crops. For local wheat, the highest-cost district is Larkana (Rs. 38), followed by Nawabshah (Rs. 37) and Jacobabad (Rs. 35). The highest AVC of Mexi-Pak wheat is in Nawabshah (Rs. 123), followed by Hyderabad (Rs. 120) and Jacobabad (Rs. 35). In the two districts, Nawabshah and Hyderabad, where Mexi-Pak is grown more widely, the AVC of this variety is considerably higher than that of local wheat. For IRRI rice, the highest AVC is reported in Larkana (Rs. 107), followed by Hyderabad (Rs. 87) and Jacobabad (Rs. 86). The AVC of local rice in Hyderabad is the same (Rs. 87), but in Larkana and Jacobabad it is lower (Rs. 58 and Rs. 50, respectively). The AVC of IRRI rice in both Larkana and Jacobabad is higher than that of local and Basmati rice. In Hyderabad and Nawabshah, the AVC of improved cotton is higher than that of local varieties; and for both the AVC per acre is higher in Nawabshah. However, the cost differential is not significant in the two districts. The AVC of sugarcane, a crop reported only in Hyderabad and Nawabshah, is Rs. 529 and Rs. 715, respectively.

AVERAGE COST OF FACTOR INPUTS BY CROP

Since one of the objectives of this study is to analyze the use of new inputs on crops, it is imperative to examine the magnitude and distribution of input costs for the various farm sizes and to compare the structure of these costs for the major crops. It also is necessary to determine the cost of inputs per acre for various crops. One note of caution here: Since the canal water costs are not included in Sind, the cost structure by inputs and the distribution of shares of inputs are somewhat distorted. Certainly the percentage share of the cost of inputs for each crop in Sind would change if the cost of canal water by individual crops was included.

Distribution of Average Cost of Factor Inputs, by Crop

The distribution of total farm costs, by input, for major crops differs a great deal between the two provinces. The share of each input is given in Tables 8.3 and A.11.

Seed Cost

The share of seed in total farm costs in Sind is higher than in the Punjab for almost every crop except sugarcane. For local wheat, it is about 82 percent in Sind and 23 percent in the Punjab. For all varieties of rice, the share of seed in the Punjab does not exceed 4 percent; but in Sind it is 43 percent for local, 37 percent for Basmati, and 15 percent for IRRI varieties. The share of cotton seed in the Punjab varies, but in Sind it constitutes 11 percent of the cost of inputs. In the Punjab the share of sugarcane seed is 36 percent, and in Sind 27 percent.

In the Punjab the share of seed in the input cost is highest in Jhelum. Rahimyar Khan also reports a rather high share for seeds in most crops. Gujranwala and Sahiwal have lower seed costs, except for wheat. Lyallpur in general reports a very low share, except for sugarcane. For almost every crop in the sample districts of the Punjab, the share of seed in the cost of inputs declines as the size of farm increases.

In Sind, where the percentage share of seed in the total input cost is generally much higher than in the Punjab, the highest percentage is reported in Jacobabad. Hyderabad is the only district where shares are reasonably low. In Sind the share of seed in total costs does not show a clear trend as the size of farm increases.

In both provinces the share of seed in the input costs of Mexi-Pak wheat and IRRI rice is less than in those of the local varieties.

TABLE 8.3

Distribution of Average Variable Cost, by Factor Inputs on Crops
(rupees)

Crop	Seed	Water	Fertilizer	Hired Labor	Farm Machinery	Marketing	Other
				Jhelum			
Wheat							
Local	161.2	32.0	55.0	166.4	134.2		48.0
	(61.2)	(0.7)	(1.2)	(17.5)	(18.5)		(1.0)
Mexi-Pak	287.5		70.0	1,193.9	190.0		
	(18.5)		(2.3)	(67.0)	(12.2)		
Rice							
Local							
Basmati							
IRRI							
Cotton							
Local							
Improved							
Sugarcane							
Maize							
Local	8.0	7.6					
	(51.3)	(48.7)					
Improved							
				Gujranwala			
Wheat							
Local	144.1	59.1	361.4	216.9	275.0		
	(17.8)	(12.9)	(34.7)	(27.1)	(7.5)		
Mexi-Pak	375.3	221.0	816.3	437.7	1,177.4	843.8	
	(18.2)	(12.6)	(41.2)	(16.3)	(10.0)	(1.8)	
Rice							
Local	17.3	101.7	425.5	272.1	504.6	206.3	
	(1.8)	(19.7)	(44.0)	(23.1)	(9.1)	(2.2)	
Basmati	29.1	162.0	414.6	320.3	361.5	367.5	
	(2.9)	(19.6)	(45.6)	(23.9)	(6.3)	(1.7)	
IRRI	17.2	150.4	455.6	429.1	729.1	593.0	
	(1.4)	(17.8)	(42.1)	(23.9)	(12.1)	(2.7)	
Cotton							
Local	22.7	64.6	205.2	71.0			
	(6.0)	(20.8)	(46.8)	(26.4)			
Improved							
Sugarcane	155.3	29.4	87.9	226.6	62.5	27.5	
	(29.4)	(7.7)	(14.6)	(45.7)	(2.3)	(0.3)	
Maize							
Local	8.0	9.7		100.0			
	(6.3)	(15.4)		(78.4)			
Improved							

(continued)

TABLE 8.3 (continued)

Crop	Seed	Water	Fertilizer	Hired Labor	Farm Machinery	Marketing	Other
				Sahiwal			
Wheat							
Local	90.4	43.9	151.8	59.0	92.3	52.5	92.0
	(20.4)	(15.3)	(44.9)	(11.3)	(7.4)	(0.2)	(0.6)
Mexi-Pak	311.2	141.1	617.7	113.5	538.7	403.1	77.8
	(16.7)	(13.0)	(38.4)	(5.0)	(11.3)	(15.3)	(0.3)
Rice							
Local	10.5	59.1	190.9	46.0	177.0	135.5	10.0
	(2.0)	(20.2)	(45.4)	(5.8)	(10.4)	(16.0)	(0.1)
Basmati	29.1	96.6	368.8	84.6	246.3	205.5	41.0
	(3.3)	(18.9)	(45.3)	(6.9)	(11.5)	(13.8)	(0.3)
IRRI	10.7	56.2	263.6	68.0	184.6	233.7	
	(1.6)	(17.2)	(44.9)	(7.6)	(7.7)	(20.9)	
Cotton							
Local	21.6	90.6	343.8	175.5	441.4	97.8	40.0
	(3.5)	(23.4)	(14.9)	(36.3)	(8.6)	(11.6)	(1.7)
Improved	34.4	139.6	437.5	188.3	446.1	239.4	58.4
	(2.8)	(21.4)	(30.6)	(19.4)	(15.1)	(10.4)	(0.3)
Sugarcane	390.2	70.0	208.4	179.7	150.3	103.0	36.0
	(38.0)	(11.8)	(21.9)	(16.6)	(5.6)	(5.8)	(0.2)
Maize							
Local							
Improved	28.1	35.6	105.4	60.2	135.3	103.8	25.6
	(8.1)	(17.5)	(30.9)	(4.7)	(15.7)	(22.6)	(0.5)
				Lyallpur			
Wheat							
Local							
Mexi-Pak	356.9	198.6	1,306.9	336.5	1,710.0	697.4	
	(11.9)	(7.7)	(46.1)	(10.4)	(16.7)	(7.1)	
Rice							
Local	2.1	14.6	73.3	92.3			
	(1.7)	(11.5)	(34.7)	(52.1)			
Basmati	8.2	40.6	181.0	150.2	148.9	60.0	
	(1.9)	(13.6)	(39.1)	(28.0)	(15.8)	(1.6)	
IRRI							
Cotton							
Local							
Improved	73.4	215.1	564.3	499.0	1,397.8	1,348.2	
	(3.2)	(11.5)	(29.5)	(26.7)	(17.0)	(12.1)	
Sugarcane	833.5	194.3	508.1	538.0	858.5	626.5	
	(32.2)	(9.8)	(24.7)	(15.0)	(9.7)	(8.5)	
Maize							
Local	3.1	6.6	55.0	25.0			
	(9.3)	(20.1)	(41.9)	(28.6)			
Improved	10.4	33.2	112.5	156.2	158.2	115.9	
	(3.6)	(9.5)	(35.3)	(32.0)	(12.5)	(7.1)	

TABLE 8.3 (continued)

Crop	Seed	Water	Fertilizer	Hired Labor	Farm Machinery	Marketing	Other
			Rahimyar Khan				
Wheat							
Local	105.9	133.8	313.1	225.9	30.0		
	(17.0)	(23.7)	(46.1)	(12.4)	(0.8)		
Mexi-Pak	218.7	175.2	936.0	641.6	1,215.0	70.0	
	(9.9)	(10.7)	(51.3)	(15.6)	(12.4)	(0.1)	
Rice							
Local	98.3	120.0	220.0	53.0	72.0		
	(32.5)	(37.8)	(10.4)	(12.5)	(6.8)		
Basmati	30.0	28.8	110.0	60.0			
	(13.1)	(12.6)	(48.1)	(26.2)			
IRRI	10.0	7.5	55.0				
	(12.5)	(19.0)	(68.6)				
Cotton							
Local	91.5	249.9	1,275.4	602.3	1,853.2	43.3	
	(5.6)	(24.5)	(52.3)	(7.9)	(9.0)	(0.8)	
Improved	95.5	212.6	3,315.8	622.0	1,642.4	137.2	
	(3.0)	(13.7)	(64.7)	(8.0)	(10.2)	(0.4)	
Sugarcane	1,019.0	158.6	1,152.9	202.0	240.4	22.2	
	(43.5)	(9.9)	(40.1)	(4.3)	(2.1)	(0.1)	
Maize							
Local	2.6	4.0	55.0		24.0		
	(10.8)	(8.2)	(56.4)		(24.6)		
Improved							
			Punjab				
Wheat							
Local	107.3	70.4	182.1	131.7	120.7	52.5	77.3
	(22.5)	(15.9)	(40.5)	(14.0)	(6.6)	(0.1)	(0.4)
Mexi-Pak	326.3	164.3	1,042.0	315.4	970.6	485.5	77.8
	(14.3)	(10.5)	(43.4)	(10.5)	(13.4)	(7.8)	(0.1)
Rice							
Local	24.6	73.5	251.2	163.2	487.9	145.6	10.0
	(3.9)	(20.9)	(42.0)	(17.3)	(9.2)	(6.5)	
Basmati	27.2	114.3	373.9	188.3	246.5	207.7	41.0
	(3.1)	(18.9)	(45.1)	(14.2)	(9.8)	(8.7)	(0.2)
IRRI	15.7	106.2	381.5	355.4	532.8	323.5	
	(1.4)	(17.7)	(42.5)	(21.7)	(11.5)	(5.1)	
Cotton							
Local	52.3	160.5	868.9	305.5	1,114.0	83.3	40.0
	(5.0)	(24.1)	(41.7)	(16.4)	(8.6)	(3.7)	(0.5)
Improved	59.3	167.0	943.0	401.3	889.1	503.5	58.4
	(3.1)	(14.8)	(38.9)	(19.8)	(14.7)	(8.6)	(0.1)
Sugarcane	627.6	105.2	465.6	340.0	409.6	271.8	36.0
	(35.8)	(10.2)	(26.9)	(14.0)	(7.0)	(6.0)	(0.1)
Maize							
Local	3.7	7.7	55.0	43.8	24.0		
	(11.4)	(18.4)	(31.8)	(33.7)	(4.6)		
Improved	19.4	32.4	101.3	125.4	143.7	106.0	25.6
	(6.1)	(13.9)	(32.8)	(16.9)	(14.3)	(15.7)	(0.3)

(continued)

TABLE 8.3 (continued)

Crop	Seed	Water	Fertilizer	Hired Labor	Farm Machinery	Marketing	Other
			Jacobabad				
Wheat							
Local	374.7	85.7		44.8	240.0	34.3	
	(82.4)	(4.7)		(2.1)	(3.8)	(7.0)	
Mexi-Pak	396.4	139.5		108.0	600.0	56.5	
	(75.4)	(8.4)		(3.1)	(5.4)	(7.7)	
Rice							
Local	729.1	419.7	707.8	249.7	1,301.3	256.0	
	(45.9)	(3.8)	(26.3)	(2.1)	(5.9)	(15.7)	
Basmati							
IRRI	773.8	494.7	954.3	395.3	1,202.1	297.6	80.0
	(28.1)	(3.7)	(30.8)	(1.9)	(24.4)	(10.8)	(0.3)
Cotton							
Local							
Improved							
Sugarcane							
Maize							
Local							
Improved							
			Larkana				
Wheat							
Local	558.1	125.0	440.0			42.1	
	(83.7)	(1.0)	(10.2)			(5.2)	
Mexi-Pak							
Rice							
Local	47.9		110.0			20.0	
	(27.7)		(63.6)			(8.7)	
Basmati	75.0		137.5			34.3	
	(36.5)		(53.5)			(10.0)	
IRRI	406.3		2,353.9		3,510.0	458.1	
	(12.4)		(71.9)		(2.3)	(13.4)	
Cotton							
Local							
Improved							
Sugarcane							
Maize							
Local							
Improved							

TABLE 8.3 (continued)

Crop	Seed	Water	Fertilizer	Hired Labor	Farm Machinery	Marketing	Other
			Nawabshah				
Wheat							
Local	470.9	240.0	460.0			60.6	
	(78.4)	(3.1)	(8.8)			(9.7)	
Mexi-Pak	510.1	90.0	1,778.5		1,680.0	308.7	
	(18.8)	(0.7)	(65.6)		(3.4)	(11.4)	
Rice							
Local							
Basmati							
IRRI							
Cotton							
Local	122.8		724.7			68.9	
	(13.4)		(79.1)			(7.5)	
Improved	261.3	172.9	1,731.7		2,520.0	176.6	
	(11.6)	(1.0)	(76.8)		(2.9)	(7.8)	
Sugarcane	4,121.5	211.7	2,579.6		3,500.0	1,062.8	
	(23.0)	(0.2)	(14.4)		(0.6)	(61.8)	
Maize							
Local	85.0	60.0	275.0			110.0	
	(14.4)	(20.3)	(46.6)			(18.6)	
Improved							
			Hyderabad				
Wheat							
Local							
Mexi-Pak	305.6	29.7	753.5	420.3	823.3	124.4	91.7
	(22.0)	(0.6)	(53.0)	(10.4)	(5.2)	(8.4)	(0.3)
Rice							
Local	68.1		312.7	122.5	600.0	49.9	
	(11.2)		(51.6)	(18.0)	(11.0)	(8.2)	
Basmati							
IRRI	227.6		642.3	434.6	1,200.0	172.7	
	(16.3)		(54.7)	(12.1)	(4.4)	(12.4)	
Cotton							
Local	50.6		852.8	430.8	240.0	68.6	
	(8.6)		(61.2)	(15.6)	(4.1)	(10.5)	
Improved	117.5	24.6	737.5	435.4	903.0	168.9	50.0
	(8.9)	(0.8)	(56.1)	(12.7)	(8.7)	(12.7)	(0.2)
Sugarcane	1,323.1	10.6	361.5	257.7	397.5	1,038.5	
	(46.7)	(0.1)	(13.4)	(4.7)	(1.5)	(33.7)	
Maize							
Local	15.0					14.5	
	(50.8)					(49.2)	
Improved							

(continued)

TABLE 8.3 (continued)

Crop	Seed	Water	Fertilizer	Hired Labor	Farm Machinery	Marketing	Other
				Sind			
Wheat							
Local	496.9	191.3	446.7	44.8	240.0	47.3	
	(81.8)	(2.1)	(8.4)	(0.3)	(0.5)	(6.9)	
Mexi-Pak	357.8	60.8	910.1	360.8	902.5	148.1	91.7
	(24.8)	(1.2)	(53.2)	(6.6)	(4.7)	(9.3)	(0.2)
Rice							
Local	577.0	419.7	555.3	201.5	1,185.0	209.4	64.0
	(43.2)	(3.5)	(28.5)	(3.3)	(6.2)	(15.1)	(0.2)
Basmati	75.0		137.5			34.3	
	(36.5)		(53.5)			(10.0)	
IRRI	367.8	455.9	1,296.0	388.5	1,544.2	322.9	80.0
	(15.0)	(0.4)	(63.5)	(3.0)	(5.1)	(12.9)	(0.2)
Cotton							
Local	80.3		740.6	430.8	240.0	68.7	
	(11.1)		(70.5)	(7.4)	(2.0)	(9.0)	
Improved	189.4	59.5	1,152.3	435.4	1,172.5	172.8	50.0
	(10.6)	(0.9)	(69.1)	(4.7)	(5.0)	(9.6)	(0.1)
Sugarcane	2,621.0	101.2	1,387.6	257.7	1,018.0	5,898.8	
	(26.7)	(0.2)	(14.2)	(0.7)	(0.8)	(57.4)	
Maize							
Local	38.3	60.0	275.0			46.3	
	(17.7)	(18.5)	(42.4)			(21.4)	
Improved							

Notes: The figures in parentheses are percentage shares of each input in the total input costs for each crop.

Blank spaces indicate data not applicable.

Source: Compiled by the author.

Water Cost

In Sind, since the cost of canal water is not included in the data, it is quite difficult to compare the percentage share of water in the total cost of inputs for the various crops. Limiting the analysis to the Punjab, the highest share goes to local cotton, followed by local rice, Basmati rice, and IRRI rice.

In general, Rahimyar Khan reports the highest share of water in the total farm costs; the lowest is in Jhelum. Among the irrigated districts, Lyallpur has the lowest. Water cost in Gujranwala, Sahiwal, and Lyallpur does not exceed 23 percent. For Mexi-Pak and IRRI rice the share of water in total costs is lower than for local varieties. There is no clear trend in the share of water cost as the farm size increases. In Sahiwal, for almost every crop the share of water goes up with the size of farm; but in Lyallpur, except for sugarcane, the share goes down with the increased farm size.

Fertilizer Cost

The share of fertilizer in the total input costs in the Punjab is lower than in Sind except for local wheat, local rice, sugarcane, and local maize. For cotton the percentage share of fertilizer in Sind is significantly higher. The reason for this is mainly the exclusion of canal water costs. Although in Sind the share of fertilizer in total costs varies from crop to crop rather considerably, in the Punjab there is some uniformity among wheat, rice, and cotton: the variation is between 39 percent and 45 percent. In both provinces the share of fertilizer for Mexi-Pak wheat and IRRI rice is higher than for local wheat and rice.

In the Punjab, Jhelum is the only district where fertilizer costs are insignificant. In general, the share of fertilizer in total costs in Rahimyar Khan is higher than in any other district. In Gujranwala and Sahiwal, the percentages for different crops are quite similar. In every district, and for almost every crop, the share of fertilizer tends to decline as the size of farm increases. The few exceptions to this general trend are in Sahiwal.

In Sind, where the percentage share of fertilizer in the total costs is higher for most crops than in the Punjab, there are great variations among crops and districts. In Hyderabad, for all crops except sugarcane, the share of fertilizer is uniformly high. The lowest share is for local wheat in Larkana and Nawabshah, and the low share in sugarcane is in Nawabshah and Hyderabad. The share of fertilizer does not appear to be affected by or related to the size of farm. For instance, in Hyderabad, while for some crops the share increases with the farm size, in others the opposite is true.

The share of fertilizer in the input costs of Mexi-Pak wheat and IRRI rice is greater than in local wheat and rice.

Hired Labor Cost

Hiring of agricultural labor is more pronounced in the Punjab. The labor costs for most crops constitute 11-22 percent of the total input costs. In Sind, on the other hand, the maximum share does not exceed 7 percent (Mexi-Pak wheat and local cotton). In the Punjab, while labor on Mexi-Pak wheat has a lower share in the total costs of this crop than is the case for local wheat, the share of labor in IRRI rice is greater than in local rice. The greater use of hired labor in the Punjab results mainly from the greater number of crops cultivated, the greater nonfarm employment opportunities for family members, and the more limited existence of tenant farming.

In the Punjab, Gujranwala is the only district where the labor costs have a uniformly high share in the cost of major crops, except for sugarcane. In Sahiwal, although labor costs do not have as high a share, they are reasonably uniform among different crops. In Lyallpur and Rahimyar Khan, the share of labor in total costs varies from crop to crop quite considerably. The share of labor seems to have a definite relationship to the size of farm only in Sahiwal, where it decreases as the size of farm increases. In general, it seems that the same trend exists in other districts.

In Sind there is no hired labor reported in Larkana and Nawabshah. In Jacobabad the hiring of labor is reported only by farmers holding over 50 acres. In Hyderabad there is some hiring of labor to work on farms of various sizes, but for no crop does the percentage share of labor in total costs exceed 18 percent. Also, in this district the share of labor in total costs changes differently by crop as the farm size increases: there are three crops for which it goes up with the farm size, while for IRRI rice it remains stable.

Farm Machinery Cost

Like hired labor, the use of farm machinery is very limited in Sind. In fact, even in the two sample districts (Jacobabad and Hyderabad) where there is some use of farm machinery, it is confined mainly to farms of over 50 acres. In the Punjab the use of farm machinery is more widespread, so the cost of farm machinery is an important component of the total input costs for most crops, ranging from 7 percent on local wheat and sugarcane to 15 percent on improved cotton. In almost every district except Jhelum, the share of farm machinery in total costs of Mexi-Pak wheat exceeds that of local wheat. Similarly, for IRRI rice the share is higher than for local and Basmati rice. The difference between improved and local cotton is still greater. For almost every crop in the Punjab, the share of farm machinery in total costs increases with the size of farm. Lyallpur reports the highest share, and for most crops Rahimyar Khan appears to have the lowest percentage share of farm machinery in total costs.

Distribution of Average Cost of Factor Inputs per Acre, by Crop

The individual input cost per acre for major crops, given in Tables 8.4 and A.12, differs greatly between the provinces, districts, and farm categories.

Seed Cost per Acre

The average cost of seed per acre for all crops in Sind is higher than in the Punjab; this is true especially of all rice varieties, sugarcane, and cotton. The lowest average cost is reported for rice in the Punjab and for cotton in Sind. In both provinces the seed cost of IRRI rice is lower than of the local and Basmati varieties. In the Punjab the cost of Mexi-Pak wheat seed (Rs. 23) is slightly higher than of local wheat (Rs. 22), while in Sind the average cost of Mexi-Pak seed (Rs. 26) is lower than of local wheat (Rs. 31). Although there is a significant cost differential in cotton varieties between the Punjab and Sind, the seed cost of improved and local cotton differs very little. The cost of sugarcane seed in the Punjab is Rs. 142 and Rs. 181 in Sind.

In the Punjab the highest average cost of seed for all crops is reported in Rahimyar Khan. The cost of wheat seed does not show a great variation between the districts and within each district for the Mexi-Pak and local varieties. While in Gujranwala, Sahiwal, and Lyallpur there is very little difference in the seed cost per acre of all rice varieties, the cost in Rahimyar Khan is much higher. In sugarcane the cost of seed is Rs. 205 in Rahimyar Khan, Rs. 128 in Sahiwal and Lyallpur, and Rs. 155 in Gujranwala.

The average cost of seed of major crops in the Punjab does not show a clear relationship with the change in size of farm; there are both increasing and declining trends. There are also cases in which the average cost of seed remains stable as the size of farm changes.

In Sind there is no single district where the average cost of seed of all crops is generally the highest. For wheat, in almost all districts the average cost remains the same: Rs. 29 for local wheat and Rs. 26 for Mexi-Pak wheat. For rice the highest average cost

TABLE 8.4

Distribution of Average Variable Cost per Acre, by Factor Inputs on Crops (rupees)

Crop	Seed	Water	Fertilizer	Hired Labor	Farm Machinery	Marketing	Other
				Jhelum			
Wheat							
Local	19.4	3.9	6.6	20.0	16.2	0.0	5.8
Mexi-Pak	14.4	0.0	3.5	59.7	9.5	0.0	0.0
Rice							
Local							
Basmati							
IRRI							
Cotton							
Local							
Improved							
Sugarcane							
Maize							
Local	42.1	40.0	0.0	0.0	0.0	0.0	0.0
Improved							
				Gujranwala			
Wheat							
Local	24.9	10.2	62.5	37.5	47.6	0.0	0.0
Mexi-Pak	23.2	13.7	50.5	27.1	72.9	52.2	0.0
Rice							
Local	2.3	13.3	55.8	35.7	66.1	27.0	0.0
Basmati	3.6	19.7	50.5	39.0	44.0	44.8	0.0
IRRI	1.7	15.1	45.7	43.0	73.1	59.5	0.0
Cotton							
Local	7.2	20.4	64.7	22.4	0.0	0.0	0.0
Improved							
Sugarcane	155.3	28.4	87.9	226.6	62.5	27.5	0.0
Maize							
Local	8.0	9.7	0.0	100.0	0.0	0.0	0.0
Improved							

TABLE 8.4 (continued)

Crop	Seed	Water	Fertilizer	Hired Labor	Farm Machinery	Marketing	Other
				Sahiwal			
Wheat							
Local	24.6	11.9	41.3	16.0	25.1	14.3	25.0
Mexi-Pak	24.2	11.0	48.0	8.8	41.9	31.4	6.1
Rice							
Local	2.4	13.5	43.7	10.5	40.5	31.0	2.3
Basmati	4.4	14.5	55.4	12.7	37.0	30.9	6.2
IRRI	2.4	12.5	58.8	15.2	41.2	52.2	0.0
Cotton							
Local	3.6	15.1	57.3	29.2	73.6	16.3	6.7
Improved	3.4	13.6	42.6	18.3	43.4	23.3	5.7
Sugarcane	128.4	23.0	68.5	59.1	49.4	33.9	11.8
Maize							
Local							
Improved	8.1	10.2	30.3	17.3	38.9	29.8	7.3
				Lyallpur			
Wheat							
Local							
Mexi-Pak	21.9	12.2	80.1	20.6	104.8	42.8	0.0
Rice							
Local	2.6	17.8	89.4	112.6	0.0	0.0	0.0
Basmati	3.5	17.1	76.1	63.1	62.6	25.2	0.0
IRRI							
Cotton							
Local							
Improved	5.8	17.0	44.5	39.3	110.2	106.2	0.0
Sugarcane	128.6	30.0	78.4	83.0	32.5	96.7	0.0
Maize							
Local	4.4	9.6	79.7	36.2	0.0	0.0	0.0
Improved	4.9	16.0	52.8	73.4	74.3	54.4	0.0
				Rahimyar Khan			
Wheat							
Local	20.2	25.5	59.6	43.0	5.7	0.0	0.0
Mexi-Pak	20.2	16.2	86.3	59.2	112.1	6.5	0.0
Rice							
Local	25.5	31.1	57.0	13.7	18.7	0.0	0.0
Basmati	15.0	14.4	55.0	30.0	0.0	0.0	0.0
IRRI	20.0	14.9	110.0	0.0	0.0	0.0	0.0
Cotton							
Local	8.2	22.4	114.1	53.9	165.8	3.9	0.0
Improved	7.0	15.7	244.2	45.8	120.9	10.1	0.0
Sugarcane	205.0	31.9	232.0	40.7	48.4	4.5	0.0
Maize							
Local	3.5	5.3	73.3	0.0	32.0	0.0	0.0
Improved							

(continued)

TABLE 8.4 (continued)

Crop	Seed	Water	Fertilizer	Hired Labor	Farm Machinery	Marketing	Other
				Punjab			
Wheat							
Local	22.4	14.7	38.0	27.5	25.2	11.0	16.1
Mexi-Pak	22.8	11.5	72.7	22.0	67.7	33.9	5.4
Rice							
Local	4.7	14.2	48.5	31.5	94.2	28.1	1.9
Basmati	4.0	17.0	55.5	27.9	36.6	30.8	6.1
IRRI	1.8	12.3	44.1	41.0	61.5	37.4	0.0
Cotton							
Local	6.5	20.1	108.6	38.2	139.3	10.4	5.0
Improved	5.1	14.2	80.3	34.2	75.7	42.9	5.0
Sugarcane	142.0	23.8	105.3	76.9	92.7	61.5	8.1
Maize							
Local	5.7	11.8	84.6	67.3	36.9	0.0	0.0
Improved	6.9	11.5	35.9	44.5	51.0	37.6	9.1
				Jacobabad			
Wheat							
Local	29.1	6.7	0.0	3.5	18.7	2.7	0.0
Mexi-Pak	26.3	9.3	0.0	7.2	39.8	3.8	0.0
Rice							
Local	22.7	13.1	22.1	7.8	40.6	8.0	2.0
Basmati							
IRRI	24.3	15.5	29.9	12.4	37.7	9.3	2.5
Cotton							
Local							
Improved							
Sugarcane							
Maize							
Local							
Improved							
				Larkana			
Wheat							
Local	32.0	7.2	25.2	0.0	0.0	2.4	0.0
Mexi-Pak							
Rice							
Local	16.0	0.0	36.7	0.0	0.0	6.7	0.0
Basmati	22.1						
IRRI	13.3						
Cotton							
Local							
Improved	22.1	0.0	40.4	0.0	0.0	10.1	0.0
Sugarcane	13.3	0.0	77.2	0.0	115.1	15.0	0.0
Maize							
Local							
Improved							

TABLE 8.4 (continued)

Crop	Seed	Water	Fertilizer	Hired Labor	Farm Machinery	Marketing	Other
				Nawabshah			
Wheat							
Local	28.9	14.7	29.2	0.0	0.0	3.7	0.0
Mexi-Pak	25.8	4.6	89.9	0.0	84.9	15.6	0.0
Rice							
Local							
Basmati							
IRRI							
Cotton							
Local	13.4	0.0	79.3	0.0	0.0	7.5	0.0
Improved	13.5	9.0	89.8	0.0	130.6	9.2	0.0
Sugarcane	164.5	8.5	102.9	0.0	139.7	441.5	0.0
Maize							
Local	8.5	6.0	27.5	0.0	0.0	11.0	0.0
Improved							
				Hyderabad			
Wheat							
Local							
Mexi-Pak	26.5	2.6	65.3	36.4	71.4	10.8	7.9
Rice							
Local	9.7	0.0	44.7	17.5	85.7	7.1	0.0
Basmati							
IRRI	14.2	0.0	40.1	27.1	74.8	10.8	0.0
Cotton							
Local	8.0	0.0	134.3	67.8	37.8	10.8	0.0
Improved	9.7	2.0	60.7	35.8	74.3	13.9	4.1
Sugarcane	246.9	2.0	67.4	48.1	74.2	193.8	0.0
Maize							
Local	15.0	0.0	0.0	0.0	0.0	14.5	0.0
Improved							
				Sind			
Wheat							
Local	30.6	11.8	27.5	2.8	14.8	2.9	0.0
Mexi-Pak	26.3	4.5	66.8	26.5	66.3	10.9	6.7
Rice							
Local	22.1	16.1	21.3	7.7	45.5	8.0	2.5
Basmati	22.1	0.0	40.4	0.0	0.0	10.1	0.0
IRRI	14.9	18.5	52.5	15.7	62.5	13.1	3.2
Cotton							
Local	10.7	0.0	98.8	57.4	32.0	9.2	0.0
Improved	12.1	3.8	73.3	27.7	74.6	11.0	3.2
Sugarcane	180.8	7.0	95.7	17.8	70.2	406.8	0.0
Maize							
Local	9.6	15.0	68.8	0.0	0.0	11.6	0.0
Improved							

Note: Blank spaces indicate data not applicable.

Source: Compiled by the author.

of seed is reported in Jacobabad and, next, in Larkana. For cotton, Nawabshah reports a higher average cost than Hyderabad, but in both districts the cost per acre of improved seed is higher than of local seed. Sugarcane, which is reported only in Nawabshah and Hyderabad, has a very high seed cost in Hyderabad (Rs. 247) as against Nawabshah (Rs. 165).

In almost every district of Sind, for most crops the average cost of seed increases with the size of farm, except for local wheat in Nawabshah and improved cotton in Hyderabad. There are only three cases in which the cost remains stable as the farm size changes.

Water Cost per Acre

Since the average cost of water in Sind does not include the cost of canal water, it would be incorrect to analyze the Sind data. In the Punjab, the average cost of water ranges from Rs. 11 (Mexi-Pak wheat) to Rs. 24 (sugarcane). The costs of water for Mexi-Pak wheat and IRRI rice are not very different. Rahimyar Khan reports the highest average cost of water per acre for most crops. The water cost in most cases does not show a clear relationship to the change in size of farm. However, where there is some trend, in general the cost tends to decline with the increase in farm size.

Fertilizer Cost per Acre

The average cost of fertilizer per acre for all crops, except for IRRI rice, is higher in the Punjab than in Sind. The cost on Mexi-Pak in both provinces is higher than on local wheat by a significant margin. The average cost of fertilizer for IRRI rice in Sind is higher than for local rice. In the Punjab, however, the fertilizer cost on IRRI rice is lower. In both provinces the fertilizer cost on improved cotton is lower than on local cotton. The cost of fertilizer on sugarcane is high in both provinces, but on local cotton the cost is higher still.

In the Punjab the highest average cost of fertilizer for most crops is reported in Rahimyar Khan, especially for improved cotton and sugarcane. Sahiwal reports the lowest average cost of fertilizer. The average cost of fertilizer for most crops in almost every sample district of the Punjab increases with the size of farm, with only few exceptions, such as Mexi-Pak wheat in Sahiwal.

In Sind, Nawabshah reports the highest average cost of fertilizer on most crops. The lowest average cost is reported in Jacobabad, for local and IRRI rice. In both Larkana and Jacobabad, the average cost of fertilizer on IRRI rice is higher than on local rice. The cost of fertilizer on local wheat is considerably lower than on Mexi-Pak wheat. The average cost of fertilizer on different crops in Sind tends to increase with the size of farm, the exceptions being Mexi-Pak wheat in Nawabshah and rice in Larkana, where the opposite seems to be true.

Hired Labor Cost per Acre

The average cost of hired labor per acre in the Punjab varies from Rs. 22 for Mexi-Pak wheat to Rs. 77 for sugarcane. However, except for sugarcane, the variation in the Punjab among other crops is not very large. In Sind the maximum average cost of labor is for local cotton (Rs. 57) and the lowest for local wheat (Rs. 3). In both provinces the average cost of labor on IRRI rice is higher than on local varieties. In the case of wheat, however, in the Punjab the cost for Mexi-Pak wheat is somewhat lower than for local wheat, whereas in Sind the cost for Mexi-Pak is relatively high. In both provinces the cost of hired labor for improved cotton is lower than for local cotton.

In the Punjab, Lyallpur and Rahimyar Khan report a high average cost of labor for most crops. Certainly in Sahiwal labor costs are low, except for sugarcane. The highest average cost of labor on sugarcane is in Gujranwala (Rs. 227), and the lowest in Rahimyar Khan (Rs. 41). The lowest cost of labor on any crop is in Sahiwal (Mexi-Pak, Rs. 29). In Gujranwala, except for sugarcane, the average cost of labor on most crops is in the range of Rs. 22-Rs. 43. In almost every district in the Punjab, the average cost of hired labor per acre decreases for most crops as the size of farm increases, especially in Sahiwal and Lyallpur.

In Sind, where hired labor is used to some extent only in Hyderabad, the average cost ranges from Rs. 18 (local rice) to Rs. 68 (local cotton). The average cost of labor on various crops in Hyderabad behaves differently with the change in size of farm. For example, for Mexi-Pak wheat, IRRI rice, and local cotton the average cost increases with the size of farm, but for sugarcane it declines.

Farm Machinery Cost per Acre

The average cost per acre of using machinery in the Punjab varies from Rs. 25 on local wheat to Rs. 139 on local cotton. In both provinces the average cost on local varieties of wheat and rice is lower than on the new seeds. Except for IRRI rice, the average cost of farm machinery is higher in the Punjab than in Sind.

In the Punjab the highest cost of machinery per acre is reported in Lyallpur. Lower average costs are reported in Sahiwal and, to some extent, in Gujranwala. The cost on Mexi-Pak varies from Rs. 42 in Sahiwal to Rs. 112 in Rahimyar Khan. For IRRI rice the average cost of machinery is Rs. 41 in Sahiwal and Rs. 73 in Gujranwala. In every district the average cost for Mexi-Pak wheat and IRRI rice is higher than for local wheat and rice. The average cost for sugarcane ranges from Rs. 48 in Rahimyar Khan to Rs. 132 in Lyallpur. The average cost for improved cotton is Rs. 43 in Sahiwal, Rs. 110 in Lyallpur, and Rs. 121 in Rahimyar Khan. The average cost for crops varies considerably with the size of farm but in different directions. In Gujranwala the cost declines with increased farm size, but in other districts, and for most crops, it increases with farm size.

In Sind, since the use of machinery is restricted to one class of farmers and mostly in Hyderabad, very little can be said. The cost in Hyderabad varies from Rs. 38 for local cotton to Rs. 75 for IRRI rice. A lower average cost is reported in Jacobabad for IRRI rice (Rs. 38). Nawabshah seems to have the highest average cost of machinery for Mexi-Pak wheat, improved cotton, and sugarcane.

CHAPTER

9

AVERAGE NET FARM INCOME AND ITS DISTRIBUTION

The average net farm income (ANFI) has been calculated by subtracting the variable cost of all crops from the value of their output. It has been determined for each sample district and, within each district, for each category of farm size. It also has been calculated for each major crop in terms of total area and on a per-acre basis.

AVERAGE NET FARM INCOME

Since the cost of producing crops in Sind, which does not include the cost of canal water, affects the average variable cost favorably, the net farm income in Sind is on the high side. It would be more realistic to view these incomes, given in Table 7.1, as somewhat overstated.

ANFI is higher in Sind (Rs. 31,978) than the Punjab (Rs. 28,202). Of all the sample districts, the highest ANFI is in Nawabshah (Rs. 60,660), followed by Lyallpur (Rs. 37,542), Larkana (Rs. 35,005), Sahiwal (Rs. 33,725), Jacobabad (Rs. 20,436), Rahimyar Khan (Rs. 18,999), Hyderabad (Rs. 18,003), Gujranwala (Rs. 15,180), and Jhelum (Rs. 2,234). Nawabshah also has the highest ANFI for each farm category; the lowest for each category is in Jhelum. In every sample district the ANFI increases with the size of farm.

In the Punjab the highest ANFI is reported in Lyallpur, followed by Sahiwal, Rahimyar Khan, Gujranwala, and Jhelum. The spread between the lowest and highest farm categories is greatest in Lyallpur (Rs. 94,000), followed by Sahiwal (Rs. 68,000), Rahimyar Khan (Rs. 50,000), Gujranwala (Rs. 32,000), and Jhelum (Rs. 3,000). For farms under 12.50 acres, the highest ANFI is in Sahiwal (Rs.

7,799), followed by Lyallpur (Rs. 6,121), Rahimyar Khan (Rs. 4,503), Gujranwala (Rs. 3,248), and Jhelum (Rs. 830). For the 12.50-25.00-acre farms, the highest is again in Sahiwal (Rs. 18,024), followed by Lyallpur (Rs. 13,821), Gujranwala (Rs. 11,328), Rahimyar Khan (Rs. 10,191), and Jhelum (Rs. 2,836). In the 25.00-50.00-acre category, the highest ANFI is reported in Lyallpur (Rs. 35,683), then Sahiwal (Rs. 34,338), Gujranwala (Rs. 21,691), Rahimyar Khan (Rs. 17,413), and Jhelum (Rs. 3,529). Over 50.00 acres, the highest ANFI is reported in Lyallpur (Rs. 101,251), followed by Sahiwal (Rs. 74,740), Rahimyar Khan (Rs. 54,558), Gujranwala (Rs. 35,529), and Jhelum (Rs. 3,800).

In Sind, Nawabshah leads in ANFI. The range between the lowest and highest farm categories is largest in Nawabshah (Rs. 129,000), followed by Larkana (Rs. 68,000), Jacobabad (Rs. 43,000), and Hyderabad (Rs. 39,000). For farms under 12.50 acres, the ANFI is highest in Nawabshah (Rs. 12,312), followed by Larkana (Rs. 6,964), Hyderabad (Rs. 4,857), and Jacobabad (Rs. 4,695). In the 12.50-25.00-acre category, the highest ANFI again is reported in Nawabshah (Rs. 29,302), then Larkana (Rs. 19,491), Hyderabad (Rs. 9,695), and Jacobabad (Rs. 9,423). For the 25.00-50.00-acre size, the ANFI is Nawabshah, Rs. 56,698; Larkana, Rs. 36,140; Jacobabad, Rs. 18,021; and Hyderabad, Rs. 15,943. Over 50.00 acres Nawabshah has the highest ANFI (Rs. 141,326), followed by Larkana (Rs. 75,090), Jacobabad (Rs. 48,293), and Hyderabad (Rs. 43,663).

AVERAGE NET FARM INCOME PER ACRE

The ANFI per acre (see Table 7.2) is higher in the Punjab (Rs. 957) than Sind (Rs. 918). However, Nawabshah leads all other districts (Rs. 1,783), followed by Lyallpur (Rs. 1,310), Sahiwal (Rs. 1,100), Larkana (Rs. 1,070), Gujranwala (Rs. 611), Jacobabad (Rs. 573), Rahimyar Khan and Hyderabad (Rs. 499), and Jhelum (Rs. 161). Except for Jhelum and Rahimyar Khan, where the ANFI per acre declines with increased farm size, the size of farm has a positive correlation with ANFI per acre. This relationship is strongest in Lyallpur and Sahiwal.

In the Punjab the highest ANFI per acre is in Lyallpur, followed by Sahiwal, Gujranwala, Rahimyar Khan, and Jhelum. The spread of ANFI per acre between the smallest and largest farms is greatest in Lyallpur (Rs. 818) and lowest in Gujranwala (Rs. 70). In Jhelum the ANFI per acre of the largest group is lower than that of the smallest group by about Rs. 102, and in Rahimyar Khan there is a negative dispersion of Rs. 50. Sahiwal leads Lyallpur in ANFI per acre for the two smaller farm sizes, but for the upper two groups

FIGURE 9.1

Average Net Farm Income per Acre, by District and Farm Size

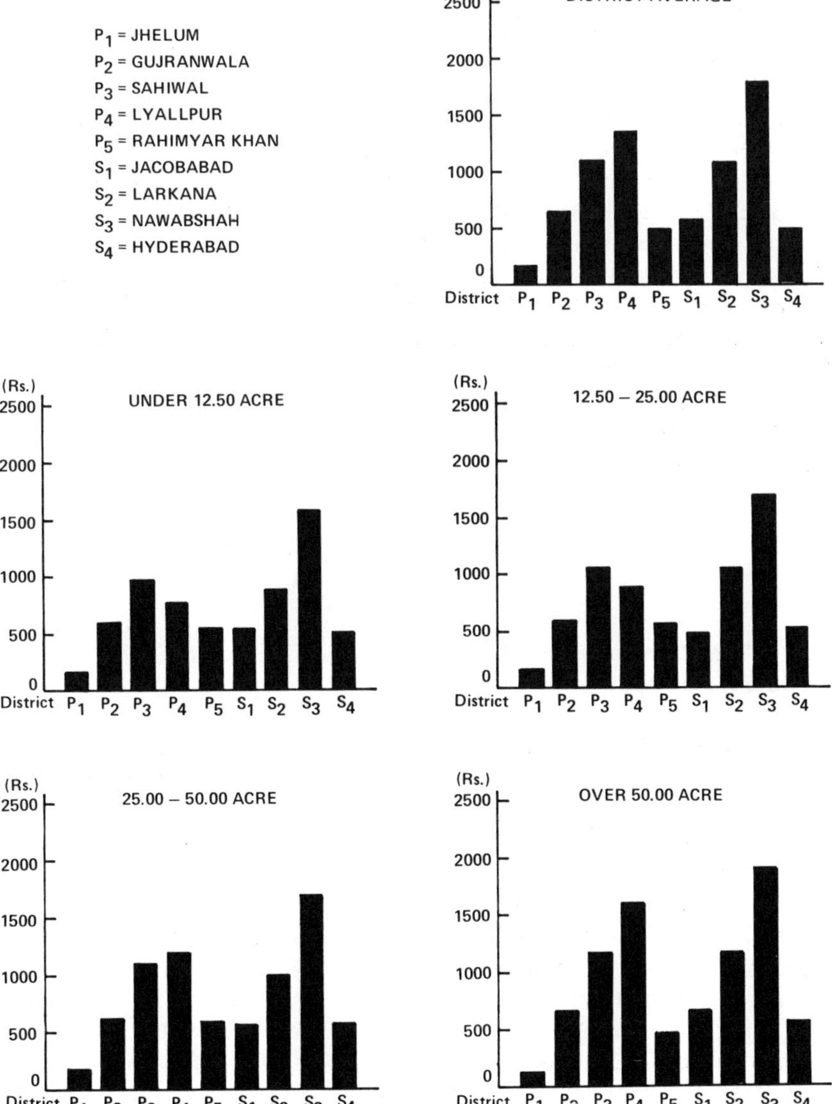

Source: Compiled by the author.

it is significantly higher in Lyallpur. Gujranwala leads Rahimyar Khan in ANFI per acre for every farm size.

The effect of farm size on ANFI per acre is seen clearly in Lyallpur, Sahiwal, and Gujranwala, where ANFI per acre increases with farm size. In Rahimyar Khan the ANFI first rises but then, for farms over 50 acres, it declines rather considerably. In Jhelum there is a consistent downward trend in ANFI per acre as the size of farm increases.

In Sind the highest ANFI is reported in Nawabshah, followed by Larkana, Jacobabad, and Hyderabad. The range between the smallest and largest farm groups is greatest in Nawabshah (Rs. 329), then Larkana (Rs. 284), Jacobabad (Rs. 120), and Hyderabad (Rs. 18). In no district is there a negative trend in the ANFI per acre, although for some reason in every district except Jacobabad, the income of farmers in the 25.00-50.00-acre category is lower than that of those with 12.50-25.00-acres. In Jacobabad the income of farmers with 12.50-25.00 acres is lower than that of those with less than 12.50 acres. ANFI per acre rises with the size of farm in every district of Sind. In almost every case the increase is largest between the 25.00-50.00-acre and over-50.00-acre sizes.

DISTRIBUTION OF AVERAGE NET FARM INCOME BY CROP

The share of major crops in the ANFI shown in Tables 9.1 and A.13 varies considerably between the two provinces. In the Punjab the major crops contribute about 83 percent to the ANFI and in Sind, about 80 percent. The picture changes a great deal with the individual crops. In wheat, while the Mexi-Pak varieties add more than do the local seeds in both provinces, it must be noted that the share of the new seeds is higher in the Punjab (17 percent) than in Sind (7 percent). The contribution of local wheat in the Punjab is insignificant, but in Sind it amounts to about 4 percent. In the case of rice, in Sind the share in the total ANFI is about 23 percent, as against 11 percent in the Punjab. While in Sind the major share is from IRRI rice (16 percent) and local rice (6 percent), in the Punjab the contribution of Basmati rice is over 8 percent, and of IRRI rice only 2 percent. The share of cotton (in both provinces improved varieties dominate) in the Punjab is about 34 percent, and in Sind 17 percent. The share of sugarcane in Sind far exceeds that in the Punjab: 29 percent versus 19 percent. In both provinces maize is an insignificant contributor to ANFI.

TABLE 9.1

Distribution of Average Net Farm Income by Crop
(rupees)

Crop	Amount	(%)	Amount	(%)	Amount	(%)	Amount	(%)
	Jhelum		Gujranwala		Sahiwal		Lyallpur	
Wheat								
Local	652.7	26.3	930.2	0.6	971.9	1.2		
Mexi-Pak	442.5	2.0	4,818.4	32.1	5,441.5	15.9	6,250.9	16.6
Rice								
Local			2,338.1	3.1	1,171.2	0.4	166.5	0.0
Basmati			4,154.2	25.6	5,607.7	14.0	1,030.6	0.5
IRRI			3,182.0	19.6	2,100.9	0.8		
Cotton								
Local			941.4	0.2	4,204.2	1.0		
Improved					13,327.4	36.3	13,919.6	36.6
Sugarcane			954.9	2.2	3,547.5	9.5	11,527.9	30.7
Maize								
Local	78.9	0.3	32.4	0.0	75.0	0.0	102.0	0.0
Improved					694.0	1.4	1,013.5	2.2
	Rahimyar Khan		Punjab		Jacobabad		Larkana	
Wheat								
Local	475.0	0.8	828.7	0.7	2,632.6	3.8	3,114.3	7.4
Mexi-Pak	1,563.6	6.5	5,082.4	16.8	2,680.2	5.8		
Rice								
Local	413.3	0.2	1,408.0	0.5	10,540.7	48.3	1,438.1	0.3
Basmati	2,021.2	0.1	4,697.6	8.6			2,411.4	0.7
IRRI	405.8	0.0	2,915.9	2.2	14,573.5	13.7	18,505.4	52.9
Cotton								
Local	5,966.3	5.9	4,669.2	1.0				
Improved	10,025.2	42.1	13,037.7	33.4				
Sugarcane	6,613.9	29.2	6,697.3	18.7				
Maize								
Local	87.6	0.0	89.5	0.0				
Improved			851.2	1.4				
	Nawabshah		Hyderabad		Sind			
Wheat								
Local	4,236.6	3.2			3,398.3	3.6		
Mexi-Pak	13,538.8	7.2	3,458.2	15.5	5,001.3	7.1		
Rice								
Local			1,048.8	0.6	8,403.2	6.4		
Basmati					2,411.4	0.2		
IRRI			5,254.8	13.5	12,693.2	16.0		
Cotton								
Local	10,221.8	2.1	4,327.5	2.9	6,754.6	1.5		
Improved	21,136.5	24.3	9,366.6	24.1	15,251.6	15.8		
Sugarcane	56,346.6	53.1	10,813.6	26.5	31,930.3	29.3		
Maize								
Local	11,410.0	0.3	560.5	0.1	4,177.0	0.2		
Improved								

Note: Blank spaces indicate data not applicable.

Source: Compiled by the author.

While in the Punjab the major crops collectively contribute about 83 percent to the total ANFI, their collective and individual shares differ widely among the sample districts and, within each district, among the various farm groups. Lyallpur leads, with about 87 percent of the ANFI contributed by the major crops, followed by Rahimyar Khan (85 percent), Gujranwala (83 percent), Sahiwal (80 percent), and Jhelum (29 percent).

In Jhelum, local wheat is the only major crop that is a significant contributor (26 percent). The share of local wheat in the ANFI increases with farm size.

Rice (over 48 percent) and wheat (33 percent) dominate in Gujranwala. In rice, Basmati (26 percent) and IRRI (20 percent) are the major contributors. The share of local wheat is negligible. The share of sugarcane in ANFI is just over 2 percent. Overall, the share of major crops has no clear relation to the size of farm: it decreases somewhat and then increases. Except for Basmati rice, in which the share tends to decline with increased farm size, the share of wheat and rice increases with farm size.

In Sahiwal, where the major crops contribute over 83 percent to the total ANFI, the position of cotton is dominant: 37 percent. Wheat (of which Mexi-Pak is the dominant variety) contributes 17 percent and rice, just over 15 percent (Basmati is the major contributor). Sugarcane accounts for 10 percent of ANFI. The share of major crops collectively tends to decline with increased farm size in Sahiwal. There is a definite increase in the share of Mexi-Pak wheat, Basmati and IRRI rice, and cotton (improved). The shares of sugarcane and local cotton decline with increased farm size.

In Lyallpur, where the major crops collectively contribute about 87 percent to ANFI, the shares of cotton (improved) and sugarcane dominate: 37 percent and 31 percent, respectively; then comes Mexi-Pak wheat with about 17 percent. Improved maize contributes about 2 percent. The collective share of these crops has a definite upward trend with increased farm size. For both cotton and sugarcane, the same strong trend is maintained; but for Mexi-Pak wheat there is an opposite trend.

Among the major crops, the position of cotton is dominant in Rahimyar Khan (48 percent); then comes sugarcane with 29 percent. For wheat the share is just over 7 percent. The improved varieties dominate both wheat and rice. Although the major crops collectively contribute about 85 percent to ANFI, their share tends to decline with increased farm size. Exceptions are cotton and sugarcane, where the upward trend is strong; but for wheat and rice the downward trend is quite noticeable.

In Sind, while the collective share of major crops in ANFI of the province is 80 percent, the collective and individual shares of these crops differ greatly among the districts and farm sizes. Nawabshah leads (90 percent), followed by Hyderabad (83 percent), Jacobabad (72 percent), and Larkana (61 percent).

In Jacobabad the position of rice (especially local varieties) is outstanding; it contributes about 62 percent to ANFI. The only other major crop is wheat, with about 10 percent (Mexi-Pak being dominant). The collective share of major crops declines as farm size increases. While the shares of local wheat and rice decline, the shares of Mexi-Pak wheat and rice increase quite clearly with the farm size.

The position of IRRI rice is most dominant (53 percent) in Larkana, the only other major crop being local wheat, with a contribution of 7 percent to ANFI. The collective share of major crops in ANFI declines with increased farm size. This is very much evident in the case of IRRI rice, but the share of local wheat remains more or less stable as farm size increases.

Sugarcane dominates in Nawabshah, with a share of 53 percent in ANFI. Then comes cotton, with over 26 percent (24 percent being contributed by the improved varieties). Wheat contributes about 10 percent, of which Mexi-Pak accounts for over 7 percent. The collective share of major crops increases with size of farm, mainly because of the share of sugarcane. This is the only district in Sind where the percentage share of major crops rises with farm size.

In Hyderabad, where the major crops collectively contribute over 83 percent to ANFI, the individual contributions of the crops are cotton and sugarcane, 27 percent; wheat, 16 percent; and rice, 14 percent. For wheat and rice, the contributions are almost exclusively by the Mexi-Pak and IRRI varieties. The collective share of major crops declines with increased farm size. The same holds true for crops individually, except for improved cotton, where the share increases with farm size.

DISTRIBUTION OF AVERAGE NET FARM INCOME PER ACRE, BY CROP

The ANFI per acre, given in Tables 9.2 and A.14, is higher for most crops in Sind. The most striking difference is in ANFI per acre of local cotton and IRRI rice. The effect of change in the size of farm on ANFI in the two provinces is neither uniform nor equally distributed among crops.

In both provinces, except in Rahimyar Khan, ANFI per acre of Mexi-Pak tends to rise with the size of farm. For IRRI rice, in every district of the Punjab and Sind (except Sahiwal), ANFI declines with increased farm size. In most cases where ANFI of a crop

TABLE 9.2

Distribution of Average Net Farm Income per Acre, by Crop
(rupees)

Crop	Jhelum	Gujranwala	Sahiwal	Lyallpur	Rahimyar Khan	Punjab	Jacobabad	Larkana	Nawabshah	Hyderabad	Sind
Wheat											
Local	78.5	160.9	264.1		90.5	173.0	204.7	178.5	259.6		209.0
Mexi-Pak	22.1	298.2	423.1	383.3	144.2	354.4	177.9		684.1	299.7	377.2
Rice											
Local		306.4	268.0	203.1	107.1	271.8	328.7	479.4		149.8	322.3
Basmati		505.1	842.0	433.0	1,010.6	697.0		709.2			709.2
IRRI		319.2	468.9		811.6	336.7	457.0	606.7		327.6	514.1
Cotton											
Local		297.0	700.7		533.7	583.7			1,118.4	681.5	900.6
Improved			1,297.7	1,096.9	738.2	1,110.5			1,095.7	770.9	970.2
Sugarcane		954.9	1,167.0	1,779.0	1,330.8	1,515.2			2,248.5	2,017.5	2,202.1
Maize											
Local	415.3	32.4	150.0	147.8	116.8	137.6			1,141.0	560.5	1,044.3
Improved			199.4	475.8		301.8					

Note: Blank spaces indicate data not applicable.

Source: Compiled by the author.

increases with farm size, the increase from the 25.00-50.00-acre farm size to that of over 50.00 acres is considerable. In both provinces, except for Jhelum in the Punjab and Jacobabad in Sind, ANFI per acre is higher for Mexi-Pak varieties than for local wheat; the most significant difference is in Nawabshah. For rice, while ANFI per acre of IRRI varieties is higher than that of local rice in every district, the income per acre from Basmati rice is higher than from IRRI rice. The ANFI per acre of improved cotton in every district (except in Nawabshah) is higher than of local cotton.

In the Punjab, ANFI per acre for most crops is highest in Sahiwal. The highest ANFI of Mexi-Pak wheat is reported in Sahiwal (Rs. 423), followed by Lyallpur (Rs. 383), Gujranwala (Rs. 298), Rahimyar Khan (Rs. 144), and Jhelum (Rs. 22). The highest ANFI per acre for IRRI rice is in Rahimyar Khan (Rs. 812), followed by Sahiwal (Rs. 469) and Gujranwala (Rs. 319). The highest ANFI for improved cotton is reported in Sahiwal (Rs. 1,298), followed by Lyallpur (Rs. 1,097) and Rahimyar Khan (Rs. 738). For sugarcane, which generally has the highest ANFI per acre among all crops (except improved cotton in Sahiwal), the highest income is reported in Lyallpur (Rs. 1,779), followed by Rahimyar Khan (Rs. 1,331), Sahiwal (Rs. 1,167), and Gujranwala (Rs. 955). In Sahiwal, Lyallpur, and Gujranwala, with a few exceptions, the ANFI per acre increases with the size of farm.

In Sind the highest ANFI per acre for every crop is in Nawabshah; the differentials are significantly high in the cases of Mexi-Pak wheat and local and improved cotton. The ANFI of Mexi-Pak wheat is highest in Nawabshah (Rs. 684), followed by Hyderabad (Rs. 300) and Jacobabad (Rs. 178). For IRRI rice the highest ANFI per acre is reported in Larkana (Rs. 607), followed by Jacobabad (Rs. 457) and Hyderabad (Rs. 328). For sugarcane, which in Sind has the highest ANFI per acre of all crops, Nawabshah leads Hyderabad (Rs. 2,249 and Rs. 2,018, respectively). The ANFI of Mexi-Pak wheat, except in Hyderabad, where it remains somewhat stable, increases with the size of farm. In both Nawabshah and Hyderabad, ANFI of sugarcane declines with increased farm size. In every district ANFI of IRRI rice declines with increased farm size. The ANFI per acre of local wheat in Jacobabad, Larkana, and Nawabshah tends to increase with farm size.

CHAPTER

10

AVERAGE FARM INCOME FROM THE SALE OF INPUTS AND OTHER SOURCES

Farm incomes are not generated by crop production alone. On most farms, income is also derived from the sale of farm inputs, renting of farm equipment, human and animal labor, and the sale of such products as milk and butter. In this study, the average farm income from sources other than crop production has been classified under the sale of human labor, the sale of animal labor, the renting of farm machinery, the sale of tubewell water, the renting of transport services, and the sale of farm products other than crop output. Using the data in Tables 10.1 and A. 15, they will be discussed.

DISTRIBUTION OF AVERAGE FARM INCOME BY FARM SIZE AND DISTRICT

The average farm income from sources other than crop output is higher in Sind (Rs. 2,517) than in the Punjab (Rs. 2,061). However, among the sample districts, Rahimyar Khan leads (Rs. 4,122), followed by Hyderabad (Rs. 3,495), Lyallpur (Rs. 3,480), Jacobabad (Rs. 3,402), Larkana (Rs. 2,387), Jhelum (Rs. 2,274), Nawabshah (Rs. 1,555), Sahiwal (Rs. 1,277), and Gujranwala (Rs. 1,223). In almost every district except Lyallpur and Nawabshah, the average income increases with the size of farm. The strongest positive relationship is evident in Larkana and Hyderabad.

DISTRIBUTION OF AVERAGE INCOME FROM THE SALE OF HUMAN LABOR

The income from human labor is significantly higher in Sind than in the Punjab: Rs. 1,910 versus Rs. 256. This is explained partly by the fact that more man-days are rented for work on or off the farm by farmers in Sind (254) than in the Punjab (43); and partly by the difference in the wage rate: Sind (Rs. 7.68) and the Punjab (Rs. 5.87).

In the Punjab, the relatively backward districts of Jhelum and Rahimyar Khan report a high number of rented man-days. Lyallpur reports no income from the sale of human labor, and Gujranwala reports only six man-days. Sahiwal reports only 24 man-days. The highest wage per man-day is reported in Gujranwala (Rs. 10.00), followed by Jhelum (Rs. 7.00), Sahiwal (Rs. 5.67), and Rahimyar Khan (Rs. 4.83).

In Sind the sale of human labor is most significant in Hyderabad (360 man-days), followed by Jacobabad (260 man-days), Larkana (235 man-days), and Nawabshah (226 man-days). The wage rate per man-day in Sind is, on the average, much higher in each district than in the Punjab. Larkana reports the highest wage rate (Rs. 8.40), followed by Nawabshah (Rs. 7.70), Jacobabad (Rs. 7.18), and Hyderabad (Rs. 6.82).

In both the Punjab and Sind, except for Larkana and Hyderabad, the incomes earned and man-days worked decline with increased farm size.

DISTRIBUTION OF AVERAGE INCOME FROM THE SALE OF ANIMAL LABOR

The income from the sale of animal labor is higher in Sind (Rs. 1,770) than in the Punjab (Rs. 152). Again, this is partly explained by the fact that more animal labor is rented in Sind (95 days) than in the Punjab (23 days). But possibly a more important reason is the differential in the rate of labor per day: Punjab, Rs. 5.89; Sind, Rs. 18.00.

In the Punjab, both Jhelum and Rahimyar Khan report no sale of animal labor. The highest rate per day is in Gujranwala (Rs. 10.00). In Sind animal labor is sold only in Jacobabad and Hyderabad (90 days and 100 days). The higher income in Hyderabad results from the high rate per day (Rs. 30.00). In both provinces there is no clear trend in the income from animal labor and the days rented with the change in farm size.

DISTRIBUTION OF AVERAGE INCOME FROM RENTAL OF FARM MACHINERY

Although more farm machinery is owned and used in the Punjab, the hours rented in Sind are more than in the Punjab (345 hours versus 182 hours). The income differential between the two provinces

TABLE 10.1

Farm Income from Sale of Inputs and Outputs Other than Crops
(rupees)

	Human Labor			Animal Labor			Farm Machinery		
District	No. of Man-Days	Rate per Man-Day (Rs.)	Avg. Income (Rs.)	No. of Days	Rate per Day (Rs.)	Avg. Income (Rs.)	No. of Hours	Rate per Hour (Rs.)	Avg. Income (Rs.)
Jhelum	235.0	7.0	1,645.0						
Gujranwala	6.0	10.0	60.0	6.0	10.0	60.0	117.9	10.0	1,179.2
Sahiwal	23.5	5.7	157.0	23.5	5.7	157.0	159.8	9.4	1,472.5
Lyallpur						233.3	213.2	8.5	2,577.5
Rahimyar Khan	332.0	4.8	1,546.5				610.0	20.0	6,700.0
Punjab	42.8	5.9	256.4	22.5	5.9	151.9	181.9	9.8	2,570.0
Jacobabad	260.0	7.2	1,760.6	90.0	6.0	540.0	540.0	24.0	1,296.0
Larkana	235.0	8.4	1,984.0						
Nawabshah	226.0	7.7	1,698.7				330.0	30.0	9,900.0
Hyderabad	360.0	6.8	2,467.6	100.0	30.0	3,000.0	300.0	16.5	4,700.0
Sind	254.0	7.7	1,910.1	95.0	18.0	1,770.0	345.0	20.0	6,943.3

Note: Blank spaces indicate data not applicable.

Source: Compiled by the author.

Tubewell Water			Transport			Avg. Income from Other Sources (Rs.)	Avg. Income from All Inputs (Rs.)
No. of Hours	Rate per Hour (Rs.)	Avg. Income (Rs.)	No. of Miles	Rate per Mile (Rs.)	Avg. Income (Rs.)		
						3,024.2	2,274.3
136.1	4.9	660.2	4,246.0	0.3	1,275.0	2,678.6	1,222.8
169.5	4.8	831.3	8,551.0	0.2	1,683.8	3,430.1	1,277.2
121.0	5.0	2,269.7	4,000.0	0.3	1,152.2	2,371.2	3,479.8
121.4	5.3	635.7				4,416.3	4,121.9
157.2	4.8	1,102.2	8,175.3	0.2	1,487.9	2,985.4	2,060.9
						3,842.9	3,401.7
						3,112.8	2,386.5
						2,349.3	1,554.5
150.0	12.0	1,050.0	150.0	3.0	450.0	5,559.3	3,494.7
150.0	12.0	1,050.0	150.0	3.0	450.0	3,851.1	2,516.5

is explained by the number of hours rented and the difference in the rate per hour: Punjab, Rs. 9.80; Sind, Rs. 20.00. However, in Sind only farmers with over 50 acres own and rent farm machinery; in the Punjab, farmers holding over 25 acres own and rent farm machinery.

In the Punjab the highest rental of farm machinery is in Rahimyar Khan (610 hours rented, Rs. 6,700 rental). Then come the districts of Lyallpur, Sahiwal, and Gujranwala. Jhelum reports no farm machinery. The higher rental in Rahimyar Khan stems mainly from the high rate per hour (Rs. 20.00), whereas it ranges from Rs. 8.50 to Rs. 10.00 in the other three districts. In all districts the renting of farm machinery is positively correlated with increase in farm size.

In Sind, Jacobabad leads Nawabshah and Hyderabad in rental income and hours rented. Larkana reports no farm machinery for rent. The rents are Jacobabad, Rs. 24; Nawabshah, Rs. 30.00; and Hyderabad, Rs. 16.50.

DISTRIBUTION OF AVERAGE INCOME FROM THE SALE OF TUBEWELL WATER

Although in Sind the use of tubewell water is very limited (the sale of tubewell water is reported only in Hyderabad), the income from this source is very close to that in the Punjab (Rs. 1,050 and Rs. 1,102). The major reason for this is the high hourly rate for water charged in Sind (Rs. 12.00) as against the Punjab (Rs. 4.80).

In the Punjab, Sahiwal leads in the sale of tubewell water (170 hours), but the income is highest in Lyallpur (Rs. 2,270). This is explained by the difference in the rate per hour, which is lower in Sahiwal than in Lyallpur. The hourly rate for water ranges from Rs. 4.79 (Sahiwal) to Rs. 5.28 (Rahimyar Khan). Except for Gujranwala, in all districts the income from the sale of tubewell water and the hours increase with the size of farm. Jhelum reports no tubewell water.

DISTRIBUTION OF AVERAGE INCOME FROM THE SALE OF TRANSPORT SERVICES

Punjab leads Sind in both the mileage covered (8,175 versus 150 miles) and the income derived from the sale of transport services (Rs. 1,488 versus Rs. 450). The higher figures for the Punjab result from the fact that both tractors and animal-driven carts are more commonly used and rented in the Punjab. However, the rate per mile charged in Sind is significantly higher (Rs. 3.00 versus Rs. 0.23 per mile). In Sind only Hyderabad reports the sale of transport, and that only for farm holdings of over 50 acres.

In the Punjab, Jhelum and Rahimyar Khan report no sale of transport services. Sahiwal leads Lyallpur and Gujranwala in both the mileage covered and the income derived: 8,551 miles and Rs. 1,684. In both Gujranwala and Lyallpur, the rate per mile is Rs. 0.30; in Sahiwal it is Rs. 0.22. In Sahiwal and Lyallpur, the income earned and the mileage covered increase with the size of farm.

DISTRIBUTION OF AVERAGE FARM INCOME FROM OTHER SOURCES

The average income from other sources is higher in Sind than in the Punjab (Rs. 3,851 versus Rs. 2,985). In every district of the Punjab and Sind, this income increases with the size of farm. The highest income is reported in Hyderabad (Rs. 5,559), followed by Rahimyar Khan (Rs. 4,416), Jacobabad (Rs. 3,843), Sahiwal (Rs. 3,430), Larkana (Rs. 3,113), Jhelum (Rs. 3,024), Gujranwala (Rs. 2,679), Lyallpur (Rs. 2,371), and Nawabshah (Rs. 2,349). It is interesting to note that Nawabshah and Lyallpur, the two relatively progressive districts, have the lowest incomes.

CHAPTER

11
FARM INDEBTEDNESS

Farm indebtedness in the survey has been evaluated by the average size of debt, the percentate of respondents in debt, the purpose for which the debt has been incurred, and the sources of debt. These indicators, shown in Tables 11.1 and A. 16, vary among the sample districts and within each district among the various categories of respondents.

PERCENTAGE OF FARMERS IN DEBT

The percentage of respondents in debt is higher in the Punjab (50 percent) than in Sind (39 percent). In the Punjab the highest percentage in debt is in Lyallpur (73 percent), followed by Jhelum (54 percent), Sahiwal (46 percent), Rahimyar Khan (44 percent), and Gujranwala (23 percent). Landless workers report the lowest percentage of debt. Although there is no clear trend of the percentage in debt as the size of farm increases, there is some indication that it is positive.

In Sind, Hyderabad has 49 percent of the respondents reporting debt, followed by Larkana (41 percent), Jacobabad (34 percent), and Nawabshah (27 percent). The highest percentage of landless workers in debt is reported in Hyderabad (36 percent). In all districts of Sind except Hyderabad, the percentage of farmers reporting debt rises as the size of farm increases.

AVERAGE SIZE OF DEBT

The size of debt is larger in the Punjab (Rs. 7,019) than in Sind (Rs. 4,730). The highest amount of debt is in Sahiwal (Rs. 9,563), and the lowest is in Jhelum (Rs. 1,311). Both Lyallpur and

TABLE 11.1

Distribution of Farm Indebtedness

District	Farmers in Debt (%)	Average Size of Debt (Rs.)	Purpose of Debt: Consumption (%)	Purpose of Debt: Production (%)	Purpose of Debt: Other (%)	Distribution of Debt by Source: Institutional: ADBP (%)	Institutional: Cooperatives (%)	Institutional: Taccavi (%)	Noninstitutional: Commission Agents (%)	Noninstitutional: Friends and Relatives (%)	Noninstitutional: Moneylenders (%)	Noninstitutional: Other (%)
Jhelum	53.6	1,310.5	6.6	67.5	24.9	13.5	0.5			86.0		
Gujranwala	22.8	6,596.2	1.0	98.0	1.0	95.7			2.5	2.6		1.0
Sahiwal	46.2	9,563.2	1.1	97.3	1.6	93.9		0.6	3.4	1.8		0.4
Lyallpur	73.0	6,323.4		96.8	3.2	72.2	4.6		11.4	10.2		1.7
Rahimyar Khan	43.6	5,907.7	12.1	87.9		56.8	1.3		1.2	32.0		8.6
Punjab	49.5	7,019.0	2.2	95.7	2.2	80.6	1.8	0.3	5.9	9.6		1.9
Jacobabad	33.9	4,405.0	8.6	57.9	33.5			11.4	57.1	2.8	28.7	
Larkana	40.7	3,070.8	4.1	76.3	33.5			6.8	60.0	0.7	32.6	
Nawabshah	27.1	2,989.0	4.4	49.8	45.8	26.4			63.0		10.6	
Hyderabad	49.1	6,256.4	16.0	84.0		79.9		2.2		6.7	10.3	0.9
Sind	39.1	4,729.8	12.0	75.2	12.9	50.6		4.1	24.0	4.6	16.3	0.6

Notes: ADBP is Agricultural Development Bank of Pakistan.
Taccavi is a government loan to farmers in distress or affected by a natural disaster.
Blank spaces indicate data not applicable.

Source: Compiled by the author.

Gujranwala in the Punjab lead Hyderabad, which reports the highest amount of debt (Rs. 6,256) in Sind. The smallest amount of debt in Sind is in Nawabshah (Rs. 2,989).

In both the Punjab and Sind, the amount of debt generally increases with the size of farm. The highest amounts are for farmers with over 50 acres: for instance, in Sahiwal it is Rs. 32,395; in Hyderabad, Rs. 31,333; and in Lyallpur, Rs. 18,000. In most districts the lowest amount of debt is reported by farmers holding under 12.50 acres. In Jhelum, on the other hand, the smallest amount of debt is incurred by farmers with over 50 acres.

PURPOSE OF DEBT

In the Punjab, about 96 percent of the farmers and landless workers are in debt for production purposes only; in Sind the figure is 75 percent. Only 2 percent of the respondents in the Punjab have used the debt for consumption, versus about 12 percent in Sind. In the Punjab, Gujranwala leads (98 percent), followed by Sahiwal and Lyallpur (97 percent), Rahimyar Khan (88 percent), and Jhelum (68 percent). In Hyderabad, 84 percent of the respondents took loans for production purposes, followed by Larkana (76 percent), Jacobabad (58 percent), and Nawabshah (50 percent).

In every district except Nawabshah, the percentage of farmers using loans for production purposes increases with the size of farm; with increased size of farm the percentage using loans for consumption declines, except in Nawabshah. In the Punjab a high percentage of landless workers took loans for consumption, while in Sind most of the landless workers took loans for production.

SOURCE OF DEBT

For analytical convenience, the sources of farm debt have been split into two groups, institutional and noninstitutional. The institutional sources are the Agricultural Development Bank of Pakistan (ADBP), cooperatives, and the taccavi loans given by the government in periods of distress. The noninstitutional sources are commission agents, friends and relatives, moneylenders, and others. The points of distinction between these two sources are the degree of regulation of debt and the interest charged.

Institutional Sources

The dominant institution of agricultural credit has for some years been the ADBP. However, it should be noted that 81 percent of the farmers and landless workers in the Punjab reported taking loans from it, and only 51 percent in Sind. In both provinces the loans from this institution have been given to farmers holding over 25 acres. In fact, in Sind only Nawabshah and Hyderabad report farmers using the ADBP loans--and all of them hold over 50 acres.

In the Punjab the highest percentage of farmers taking loans from the ADBP is in Gujranwala (96 percent), followed by Sahiwal (94 percent), Lyallpur (72 percent), Rahimyar Khan (57 percent), and Jhelum (14 percent). In almost every district, farmers holding over 25 acres have had access to these loans.

Sind reports that 4 percent of the farmers took taccavi loans from the government. In the Punjab a negligible percentage of farmers use taccavi. The highest percentage of farmers in Sind using it is in Jacobabad (over 11 percent), followed by Larkana (7 percent) and Hyderabad (2 percent). The taccavi loans have been reported only by farmers holding over 50 acres.

Very few farmers in the sample report taking loans from cooperatives: only 2 percent in the Punjab, with Lyallpur reporting 5 percent of the respondents. In Sind no respondent used the cooperatives for loans.

Noninstitutional Sources

In Sind commission agents and moneylenders are the major sources of noninstitutional credit, whereas in the Punjab friends and relatives are the main source. In Sind, 24 percent of the respondents reported taking loans from commission agents and 16 percent from moneylenders; only 6 percent in the Punjab reported loans from commission agents. Nawabshah, Larkana, and Jacobabad report 63 percent, 60 percent, and 57 percent, respectively, using commission agents. In Lyallpur, 11 percent of the respondents had used this source of credit. Moneylenders are most active in Larkana, Jacobabad, and Nawabshah. The percentage of respondents using commission agents and moneylenders for loans declines with increased farm size, for it is the landless workers and small farmers who usually depend on these two sources.

Friends and relatives are more important in the Punjab: 10 percent of the respondents took loans from them, as against only 5 percent in Sind. Jhelum leads, with 86 percent of the respondents, followed by Rahimyar Khan (32 percent), Lyallpur (10 percent), Gujranwala (3 percent), and Sahiwal (2 percent). In Sind, only in Hyderabad did respondents take loans from friends and relatives (about 7 percent). Except for Sahiwal, in every sample district of the Punjab and Sind the percentage of farmers using friends and relatives for credit declines with increased farm size. Most of the landless workers depend on this source for their credit needs.

CHAPTER 12
LANDLESS WORKERS

In the survey an attempt was made to collect information on the landless workers, those who neither own nor lease agricultural land. These workers have been classified into eight categories: salaried worker, menial servant, shopkeeper, carpenter, blacksmith, technician, schoolteacher, and other. Table 12.1 concerns the size of the household, the agricultural and nonagricultural employment, and incomes during 1972-73.

AVERAGE NUMBER OF FAMILY MEMBERS

The average family is smaller in the Punjab than in Sind. The highest number of family members is reported in Larkana (12.5), and the lowest in Hyderabad (5.8). In the Punjab, except for Jhelum (where the size is 8.7), the average is 6.5-6.7 members per household. In Sind the numbers vary considerably: Jacobabad, 9.6; Larkana, 12.5; Nawabshah, 11.3; and Hyderabad, 5.8. The range within one district is largest in Jhelum (5.5), and the smallest in Lyallpur and Nawabshah (0.5). In other districts the range varies from 1.7 to 2.9. In no category of landless workers is there a uniformly high or low number of family members.

AVERAGE AGRICULTURAL AND NONAGRICULTURAL EMPLOYMENT

Employment of landless workers in agriculture is limited in both provinces. Only in Hyderabad (171 man-days), Sahiwal (157 man-days) and Gujranwala (128 man-days) is the agriculture-related

TABLE 12.1

Family Size, Income, and Employment of Landless Workers

Type of Worker	Average Number in Family	Employment and Income: Agriculture: Man-Days	Agriculture: Average Income (Rs.)	Nonagriculture: Man-Days	Nonagriculture: Average Income (Rs.)	Total Man-Days Employed	Total Man-Days Unemployed
			Jhelum				
Salaried worker	7.0			365.0	1,950.0	365.0	
Menial servant	10.0			365.0	3,600.0		
Shopkeeper							
Carpenter	6.5			228.0	930.0	228.0	137.0
Blacksmith							
Technician	12.0			365.0	1,800.0	365.0	
Schoolteacher							
Other	8.4			264.0	692.0	264.0	101.0
All workers	8.7			317.0	1,794.0	317.0	119.0
			Gujranwala				
Salaried worker	6.0	340.0	2,400.0	200.0	1,500.0	540.0	
Menial servant	5.0			180.0	1,000.0	180.0	185.0
Shopkeeper	7.0			350.0	2,800.0	350.0	15.0
Carpenter	7.0	33.0	67.0	347.0	2,433.0	380.0	95.0
Blacksmith	7.8			250.0	2,433.0	250.0	38.0
Technician							
Schoolteacher							
Other	7.2	12.0	150.0	298.0	2,629.0	310.0	75.0
All workers	6.6	128.0	663.0	271.0	2,133.0	399.0	82.0
			Sahiwal				
Salaried worker	6.2			344.0	2,292.0	344.0	50.0
Menial servant	7.5	338.0	1,000.0	116.0	6,750.0	454.0	36.0
Shopkeeper	5.8			284.0	3,475.0	284.0	53.0
Carpenter	7.1	81.0	389.0	233.0	5,222.0	314.0	71.0
Blacksmith	6.8	52.0	275.0	319.0	3,857.0	371.0	109.0
Technician	6.0			275.0	5,250.0	275.0	90.0
Schoolteacher	6.1			365.0	3,400.0	365.0	
Other	7.8			262.0	9,445.0	262.0	90.0
All workers	6.7	157.0	555.0	288.0	4,961.0	290.0	71.0
			Lyallpur				
Salaried worker							
Menial servant	7.0		1,200.0		600.0		
Shopkeeper							
Carpenter	6.5	62.0	350.0	208.0	2,633.0	270.0	130.0
Blacksmith	6.6	40.0	500.0	227.0	1,960.0	267.0	125.0
Technician	7.0			190.0	1,800.0	190.0	175.0
Schoolteacher							
Other	6.7	74.0	463.0	202.0	2,050.0	276.0	124.0
All workers	6.7	59.0	628.0	207.0	1,809.0	266.0	139.0
			Rahimyar Khan				
Salaried worker							
Menial servant							
Shopkeeper							
Carpenter	6.4			339.0	1,724.0	339.0	25.0
Blacksmith	7.5			365.0	1,900.0	365.0	
Technician							
Schoolteacher							
Other	5.8	13.0	40.0	313.0	1,118.0	326.0	33.0
All workers	6.5	13.0	40.0	339.0	1,581.0	352.0	29.0

(continued)

TABLE 12.1 (continued)

Type of Worker	Average Number in Family	Employment and Income: Agriculture: Man-Days	Employment and Income: Agriculture: Average Income (Rs.)	Employment and Income: Nonagriculture: Man-Days	Employment and Income: Nonagriculture: Average Income (Rs.)	Total Man-Days Employed	Total Man-Days Unemployed
				Jacobabad			
Salaried worker							
Menial servant	10.1	96.0	1,529.0	104.0	574.0	200.0	165.0
Shopkeeper	9.5			365.0	3,285.0	365.0	
Carpenter	10.0			365.0	2,920.0	365.0	
Blacksmith							
Technician							
Schoolteacher							
Other	8.7			323.0	3,136.0	323.0	42.0
All workers	9.6	96.0	1,529.0	289.0	991.0	313.0	42.0
				Larkana			
Salaried worker							
Menial servant	13.6	107.0	2,787.0	33.0	570.0	140.0	
Shopkeeper	10.7			174.0	3,218.0	174.0	
Carpenter	15.0			210.0	1,680.0	210.0	
Blacksmith	11.0			285.0	2,495.0	285.0	
Technician							
Schoolteacher							
Other	12.4	20.0	380.0	185.0	2,473.0	205.0	
All workers	12.5	63.0	1,583.0	177.0	2,087.0	240.0	
				Nawabshah			
Salaried worker							
Menial servant	11.4	94.0	1,546.0	48.0	532.0	142.0	223.0
Shopkeeper	11.6			363.0	1,817.0	363.0	2.0
Carpenter							
Blacksmith							
Technician							
Schoolteacher							
Other	11.1	14.0	214.0	259.0	2,605.0	273.0	92.0
All workers	11.3	54.0	880.0	224.0	1,651.0	228.0	137.0
				Hyderabad			
Salaried worker	6.8			354.0	1,745.0	354.0	11.0
Menial servant	6.5	331.0	1,004.0		2,803.0	331.0	34.0
Shopkeeper	6.5	10.0	40.0	336.0	1,000.0	346.0	19.0
Carpenter	4.0			300.0		300.0	65.0
Blacksmith							
Technician							
Schoolteacher							
Other	4.5			326.0	1,563.0	326.0	39.0
All workers	5.8	171.0	522.0	329.0	1,778.0	331.0	34.0

Note: Blank spaces indicate data not applicable.

Source: Compiled by the author.

employment of some significance. In Jhelum and Rahimyar Khan, no or very little employment in the agricultural sector is reported. In the Punjab, carpenter and blacksmith, and in Sind, menial servant, are the dominant categories. In Sahiwal, Lyallpur, and Gujranwala (the leading agricultural districts in the Punjab), blacksmiths and carpenters dominate the employment related to agriculture.

Nonagricultural activities provide varying amounts of employment to landless workers. For instance, Rahimyar Khan reports the highest number of man-days (339), followed by Hyderabad (329), Jhelum (317), Jacobabad (289), Sahiwal (288), Gujranwala (271), Nawabshah (224), Lyallpur (207), and Larkana (177). In the Punjab, menial servants, technicians, and blacksmiths dominate the nonagricultural employment. In Sind, shopkeepers and menial servants are more visible.

The extent of unemployment in 1972-73 is measured by the number of man-days reported without work. The lowest number of unemployed man-days is in Rahimyar Khan (29) and the highest in Lyallpur (139), followed by Nawabshah (137), Jhelum (119), Gujranwala (82), Sahiwal (71), and Jacobabad (42). Again no single category of landless workers reports a uniformly high or low number of man-days unemployed.

AVERAGE AGRICULTURAL AND NONAGRICULTURAL INCOME

The average income from agricultural pursuits is, on average, higher in Sind than in the Punjab. The leading district is Larkana (Rs. 1,583), followed by Jacobabad (Rs. 1,529), Nawabshah (Rs. 880), Gujranwala (Rs. 663), Lyallpur (Rs. 628), Sahiwal (Rs. 555), Hyderabad (Rs. 522), and Rahimyar Khan (Rs. 40). The highest average agricultural income is earned by salaried workers in Gujranwala (Rs. 2,400) and by menial servants in Larkana (Rs. 2,787). Blacksmiths and carpenters report usually lower agricultural income in both provinces.

The nonagricultural income, which is the major source for landless workers in the two provinces, is in general higher in the Punjab than in Sind. The leading district is Sahiwal (Rs. 4,961), followed by Gujranwala (Rs. 2,133), Larkana (Rs. 2087), Lyallpur (Rs. 1809), Jhelum (Rs. 1,794), Hyderabad (Rs. 1,778), Nawabshah (Rs. 1,651), Rahimyar Khan (Rs. 1,581), and Jacobabad (Rs. 991). In each district of the Punjab, the highest annual income from nonagricultural sources is earned by a different category of landless workers. In Sind, except for Nawabshah, shopkeepers seem to dominate.

CHAPTER 13 HYPOTHESES TESTED

In the literature on the "Green Revolution," several hypotheses have been propounded and many generalizations made. Some of these have been tested by field studies, but many remain generalizations without adequate research and analysis. For Pakistan, it would be fair to say that empirical studies of the process of adoption of new wheat and rice seeds are based primarily on secondary data. Consequently, the analysis of the likely consequences of the "Green Revolution" has been done in the most general terms.[1] It is hoped that this study, by using primary data collected in the sample survey, will fill the gap in the literature on Pakistan's agriculture. Also, the results of this study produce several generalized propositions that will be relevant and of interest to research workers in other underdeveloped countries.

In view of the objectives stated in Chapter 1, only those hypotheses are being tested that seem central to an understanding of the implications of the "Green Revolution." The group of hypotheses tested here is, therefore, by no means exhaustive.

IS THE PARTICIPATION IN THE ADOPTION AND USE OF NEW INPUTS BY LARGE AND SMALL FARMS DIFFERENT?

One way to evaluate the differences between incomes derived from the new high-yield varieties (NHYV) on large and small farms is to analyze the level of participation by each, which is well reflected in the the adoption and the levels of use of the new seeds and other inputs.

The first observation is that the Punjab and Sind exhibit significant differences in the adoption of the new seeds. First, adoption began earlier in the Punjab and a higher percentage of farmers in this province reported using the new wheat seeds at least once. Second, in both provinces, adoption began earlier in the more progressive districts. Third, in general, in both provinces, the large farms adopted the new wheat seeds earlier than the small farms.

In the case of IRRI rice, while adoption in the Punjab and Sind started more or less at the same time, a higher percentage of farmers had used the new rice seeds in Sind. There is evidence that, except for Sahiwal in the Punjab and Larkana in Sind, adoption started earlier on the large farms; and a higher percentage of farmers on the large farms had used the new seeds at least once.

The use of the new wheat and rice seeds is reflected by the percentage of farmers using the new seeds and the percentage of crop area planted with the new seeds. In this study, the percentage of farmers using the new seeds has been calculated under two different assumptions: farmers growing any crop and farmers growing only the crop referred to.

In wheat, of the farmers growing any crop, a significantly higher percentage used the Mexi-Pak seeds in the Punjab than in Sind. Of the farmers growing only wheat, a higher percentage reported using the new seeds in the Punjab. Also, the use of Mexi-Pak seeds is more diffused in the Punjab. In both provinces, except for Rahimyar Khan, large and small farms report a uniformly distributed pattern. From the area put to the NHYV of wheat on the different sizes of farms, it appears that the large and small farms do not differ significantly. Since the distribution of farms in the Punjab (and more so in Sind) is highly skewed, the gain from the NHYV on the smaller farms is distributed among a large number and on the larger farms it is confined to relatively fewer.[2] If the NHYV of wheat are more profitable than the local varieties, it will lead to a greater disparity in farm incomes.

The chemical fertilizer used in the Punjab and Sind is mainly nitrogenous. A very high percentage of farmers, more or less the same in the two provinces, had used the nitrogenous fertilizer at least once, but adoption began earlier in the Punjab. In both the Punjab and Sind, a higher percentage of farmers on the large than on the small farms had used fertilizer at least once and had started using it earlier.

In both provinces, on average, a greater amount of fertilizer per acre was used on the NHYV of wheat and rice, compared with their local counterparts, except for Basmati rice in the Punjab. The reason for a higher dose on Basmati rice than on IRRI rice may be explained by the farmer's calculation of a greater return per acre from Basmati. Since the price per maund of Basmati was higher, and a larger amount of fertilizer was associated with higher production per acre (although actually, in all districts of the Punjab the yield per acre of Basmati was lower than of IRRI rice), the farmers had greater expectations of cash return from Basmati.

In the Punjab, though the distribution of farmers using fertilizers on the NHYV was more or less equal on the large and small farms, the average amount of fertilizer used per acre increased with the size of farm. In Sind, where the average amount of fertilizer used per acre on the NHYV of wheat and rice was greater than in the Punjab (except in Hyderabad), there was no significant difference between the large and small farms with regard to the percentage of farmers using fertilizer and the average amount used.

Although this study does not include data on the percentage of the fertilized area planted with the NHYV of wheat and rice, there is evidence that in general the small farms fertilized a lower percentage of the area under new seeds.[3] The average yield per acre of the NHYV in the Punjab is greater than in Sind, and in both provinces the yield per acre tends to increase with the size of farm, so the return on the higher fertilizer use on all farms in Sind seems to be lower. Certainly, this conclusion is warranted in the case of large farms.

In the case of pesticides, although the percentage of farmers using them is more or less the same in the Punjab and Sind, adoption began earlier in Sind. The large farms started the use of pesticides earlier, and a higher percentage of them had used this input at least once.

In the use of farm machinery (in which the tractor dominates), the process of adoption began at the same time in both provinces, although there is strong evidence that the large farms used it more than small farms. The size effect is more pronounced in Sind, especially in the case of Mexi-Pak wheat. In the Punjab, the use of farm machinery was greater and more diffused. The yield per man-day of the NHYV, which is higher than of the local varieties, appears to increase with greater use of farm machinery. This is observed particularly on the large farms, which use farm machinery to a greater extent.

The use of tubewell water (a supplemental source of irrigation) is confined mainly to the Punjab, where the relatively more progressive districts of Sahiwal and Gujranwala began its use earlier than Rahimyar Khan. The large farms report a higher percentage of farmers owning tubewells and using water from them. The amount of water used on Mexi-Pak wheat and IRRI rice is greater in the Punjab.

ARE THE NEW SEEDS MORE PROFITABLE THAN THE LOCAL SEEDS OF WHEAT AND RICE, AND IS THERE ANY DIFFERENCE IN THE PROFITABILITY ON THE LARGE AND SMALL FARMS?

To test the first part of this hypothesis, it must be assumed that the NHYV of wheat and rice have a higher yield per acre and a higher net return per acre (value of excess yield minus the additional cost per acre of the NHYV) than the local varieties. The second part of the hypothesis must assume that the proportion of crop area given to the NHYV is the same or larger on large farms than on small ones; that the average yield per acre is at least the same, or greater, on the large than on small farms; and that the additional cost incurred by the large farms to obtain the excess yield from the NHYV of wheat and rice is less than the value of excess yield. If these three assumptions hold true, then the disparity in farm incomes between the large and small farms will tend to widen.

That the new seeds of wheat and rice are more profitable than the local seeds (see Table 13.1) is reflected in the fact that their yield per acre is higher in both the Punjab and Sind and that the net return per acre from the NHYV of wheat and rice is positive (the value of excess yield is greater than the additional cost incurred). The only exception is farms over 50 acres in Rahimyar Khan. The profitability of the new seeds is higher in the Punjab, and in the more progressive districts of the two provinces. The higher profitability in the Punjab and, in each province, in the more progressive districts is explained either by the fact that the input levels used are higher or that they are combined more judiciously.

The income disparity between the large and small farms will tend to increase if it can be demonstrated that the NHYV of wheat and rice are more profitable on the large farms. To evaluate the relative profitability of the new seeds on the large and small farms, the data on the proportion of area under the new seeds, the yield per acre, and the net return per acre should be analyzed.

The proportion of area given to the NHYV of wheat and rice (higher for wheat in the Punjab and for rice in Sind) does not seem to change with increased farm size. The few exceptions are the more backward districts of Jhelum (wheat), Rahimyar Khan (rice), and Jacobabad (rice), where the proportion of area under the new seeds increases with the size of farm. There is no case in which the proportion of crop area given to the new seeds declines with increased farm size. Thus it would appear that, at least in the relatively more backward districts, the potential for higher profitability and the consequent income disparity between the large and small farms is greater.

TABLE 13.1

Additional Income per Acre from New Wheat and Rice Seeds

	Excess Yield (mds.)		Value of Excess Yield (Rs.)		Additional Cost (Rs.)		Additional Income of New Seeds (Rs.)	
	Wheat	Rice	Wheat	Rice	Wheat	Rice	Wheat	Rice
Jhelum								
< 12.50								
12.50-25.00								
25.00-50.00								
> 50.00								
Gujranwala	5.6	9.7	117.6	232.8	-13.7	0.4	131.3	232.5
< 12.50	4.9	12.2	102.9	292.8	12.7	4.3	115.6	288.5
12.50-25.00	9.2	6.7	193.2	160.8	-4.9	-8.0	198.1	168.8
25.00-50.00	4.2	9.4	88.2	225.6	-20.3	3.3	108.5	222.3
> 50.00	5.7	9.6	119.7	230.4	1.1	-7.3	118.6	237.7
Sahiwal	6.7	12.7	140.7	304.8	24.1	27.3	116.6	277.5
< 12.50	3.2	6.8	67.2	163.2	1.7	12.6	65.6	150.6
12.50-25.00	5.8	12.0	121.8	288.0	21.0	48.6	100.8	239.4
25.00-50.00	7.0	10.7	147.0	256.8	18.3	9.9	128.7	246.9
> 50.00	7.6	15.7	159.6	376.8	18.9	33.2	140.7	343.6
Lyallpur								
< 12.50								
12.50-25.00								
25.00-50.00								
> 50.00								
Rahimyar Khan	6.1	24.0	128.1	576.0	86.1	82.1	42.1	494.0
< 12.50	6.4		134.4		18.5		115.9	
12.50-25.00	5.1	24.0	107.1	576.0	35.5	71.0	71.6	505.0
25.00-50.00	4.5		94.5		44.1		50.4	
> 50.00	4.3		90.3		97.7		-7.4	
Punjab	10.2	11.8	241.2	283.2	59.7	8.9	181.5	274.3
Jacobabad		5.9		141.6		36.9		104.7
< 12.50		11.9		285.6		33.3		252.3
12.50-25.00		11.4		273.6		42.9		230.7
25.00-50.00		5.9		141.6		27.1		114.6
> 50.00		5.1		122.4		34.8		87.6
Larkana		13.9		333.6		49.7		283.9
< 12.50								
12.50-25.00								
25.00-50.00		16.0		384.0		37.4		346.7
> 50.00		13.1		314.4		52.1		262.3
Nawabshah	23.3		489.3		100.2		389.1	
< 12.50	25.9		543.9		99.7		444.2	
12.50-25.00	18.7		392.7		87.8		304.9	
25.00-50.00	20.9		438.9		100.6		338.3	
> 50.00	26.2		550.2		105.1		445.1	
Hyderabad		7.1		170.4		0.3		170.2
< 12.50		7.6		182.4		-10.4		192.8
12.50-25.00		1.2		28.8		-3.1		31.9
25.00-50.00		5.6		134.4		38.2		96.2
> 50.00		8.2		196.8		6.5		203.3
Sind	11.4	7.1	235.2	170.4	68.4	48.2	166.8	122.3

Notes: The price of wheat is Rs. 21 per maund; of rice, Rs. 24 per maund; no local wheat in Lyallpur and Hyderabad; no Mexi-Pak wheat in Larkana; no rice is reported in Jhelum, Lyallpur, and Nawabshah.

Blank spaces indicate data not applicable.

Source: Compiled by the author.

Even though the large farms do not allocate a higher proportion of the crop area to the new wheat and rice seeds, the NHYV may still be more profitable to them if their yield per acre is higher than on small farms. The yield per acre of the NHYV of wheat in the Punjab and Sind (except for Nawabshah) is greater on the large farms than on small ones. Even with the same proportion of crop area under the new seeds on the large and small farms, the fact that the yield per acre is higher on the large farms results in greater profitability and widening income disparity. In the Punjab the yield per acre increases with the size of farm in Sahiwal, but there is no discernible trend in the yield per acre in Gujranwala and Rahimyar Khan. In Sind, in all districts the yield per acre is negatively correlated with the change in size of farm in Jacobabad, Larkana, and Hyderabad. If it is true that the small farms are able to get an equal or higher yield per acre of rice, it may be that the inputs required for the NHYV are neutral to the scale of farming.

The higher yield per acre on large farms does not necessarily bring a higher net return, unless the value of excess yield of the NHYV is greater than the additional cost incurred. The large farms will incur a loss, despite a higher yield, if the additional cost exceeds the value of excess yield. Moreover, the possibility of widening income disparity will be checked if the additional cost per acre increases more than proportionately to the increase in the value of excess yield, because it will tend to decrease the net return per acre.

In every district, the value of excess yield is greater than the additional cost per acre, indicating in general the profitability of the NHYV of wheat and rice. The value of excess yield and the net return per acre of Mexi-Pak wheat in Sahiwal increase markedly with the size of farm; the same tends to be the case in Gujranwala. But in Rahimyar Khan, there is a definite decline in the value of excess yield and net return per acre with increased farm size. In Nawabshah, farm size has an erratic effect on these indicators of profitability, although in general they tend to increase with the size of farm.

In the case of IRRI rice, only in Sahiwal is there a strong and positive relationship between the size of farm and the value of excess yield and net return per acre. The same trend exists, to some extent, in Hyderabad. In Gujranwala, Jacobabad, and Larkana, the value of excess yield and net return per acre decline with increased farm size. It is safe to say that the profitability of IRRI rice on the large farms in the rice-growing areas of the Punjab and Sind is less than on small farms.

Related to the profitability of the new wheat and rice seeds is the question of economies of scale. It has been contended by some that the new technology is neutral to scale.[4] Upon examination of the average variable cost per unit of output of Mexi-Pak wheat and IRRI rice (on a per-acre basis), given in Table 13.2, the neutrality

TABLE 13.2

Average Variable Cost per Unit of Output of the New Seeds
(rupees)

District/Farm Size	Mexi-Pak Wheat	IRRI Rice
Jhelum		
Under 12.50		
12.50-25.00		
25.00-50.00		
Over 50.00		
Gujranwala		
Under 12.50	5.71	2.87
12.50-25.00	5.84	3.95
25.00-50.00	6.15	3.85
Over 50.00	7.28	4.61
Sahiwal		
Under 12.50	5.79	4.23
12.50-25.00	5.47	4.81
25.00-50.00	5.74	4.39
Over 50.00	6.33	4.77
Lyallpur		
Under 12.50	5.07	
12.50-25.00	4.97	
25.00-50.00	5.90	
Over 50.00	8.23	
Rahimyar Khan		
Under 12.50	9.17	
12.50-25.00	9.42	
25.00-50.00	11.83	
Over 50.00	14.50	
Jacobabad		
Under 12.50		2.39
12.50-25.00	4.13	3.08
25.00-50.00	3.96	2.57
Over 50.00	4.18	4.14
Larkana		
Under 12.50		2.97
12.50-25.00		3.16
25.00-50.00		2.96
Over 50.00		2.70
Nawabshah		
Under 12.50	4.00	
12.50-25.00	4.61	
25.00-50.00	4.30	
Over 50.00	3.97	
Hyderabad		
Under 12.50	4.17	2.27
12.50-25.00	5.00	3.38
25.00-50.00	5.79	4.37
Over 50.00	6.45	4.34

Note: Blank spaces indicate data not applicable.

Source: Compiled by the author.

assumption fails the test. The cost per unit of output for Mexi-Pak wheat in every district, except Jacobabad and Nawabshah, increases with the size of farm. A similar pattern emerges for IRRI rice, though not as strongly. The increased cost per unit of output can best be explained by the increased weight of such inputs as farm machinery, hired labor, and tubewell water.

ARE THE BENEFITS OF THE "GREEN REVOLUTION" UNEQUALLY DISTRIBUTED BETWEEN THE LARGE AND SMALL FARMS?

It has been demonstrated that the new wheat and rice seeds are more profitable than the local seeds. The major reason is not only the higher yield per acre of the new seeds but, more importantly, their higher net farm income per acre. Also, there is evidence (although not equally strong) that these seeds are more profitable to the large farms, at least in the more progressive districts. Now if it can be demonstrated that the large farms had adopted the new seeds and inputs earlier (new seeds being more profitable than the local seeds); had used more of the new inputs, especially those subsidized by the government; had substituted the new seeds for local seeds and the higher-value crops for lower-value crops to a greater extent; had marketed a higher proportion of the grain output, which has had price support through government procurement; and had greater and easier access to agricultural credit from the institutional sources, then it will be safe to conclude that they have been deriving greater benefits from the "Green Revolution."

In this study, it has been established (certainly in the Punjab) that the large farms began adoption of the new inputs earlier and used greater amounts of them per acre. This is true for Mexi-Pak wheat and, to some extent, for IRRI rice. Also, in both the Punjab and Sind, the large farms have substituted the new seeds for local varieties and the higher-value crops for lower-value crops a great deal more than small farms.

In Pakistan the critical inputs--chemical fertilizer, pesticides, farm machinery, and water--that have made the new seeds more profitable have been subsidized by the government for a number of years. For instance, it has been estimated that farmers were given a fertilizer subsidy of Rs. 207 million in 1972-73, of which about Rs. 183 million was spent in the Punjab and Sind. The extent of subsidy was about 35 percent on fertilizer. In the same year, the government provided a subsidy of Rs. 111 million on pesticides, of which a major part was used in these two provinces. The extent of subsidy on pesticides was 75 percent in the Punjab and 50 percent in

Sind. In addition, the government spent about Rs. 50 million on aerial spraying, which was free of cost to the farmers.[5] Although no precise estimates were available for the subsidy on farm machinery and tubewells, it is said that a sizable amount was provided to the farmers in 1972-73. It is well known that irrigation water provided by the canal system in the Indus Basin is subsidized. In view of the fact that the owners or operators of large farms have greater access to water, and often use more water per acre, the benefits of a subsidy on water would be distributed unequally between the large and small farms.[6]

Since it has been demonstrated that the large farms marketed a larger proportion of grain output (and of other crops), they must have benefited more from the price support and procurement policies of the government. Also, because of the greater waiting capacity of the large farms, the benefits of the open-market sale of grain output must have been greater to them than to the small farms.

Looking at agricultural credit, its structure, and its sources, the large farms were able to use more credit, obtained on easier terms, for the purchase of subsidized inputs. This has undoubtedly increased the ownership and use of tubewell water and farm machinery by the large farms in the Punjab and, to some extent, in Sind. From the data on agricultural credit given in this study, it is clear that the institutional sources (the Agricultural Development Bank of Pakistan is the leader), which have been the dominant sources to the large farms, have given credit to the farmers at a low rate of interest (7-8 percent).

Even if it is difficult to determine the quantum of benefits that the large and small farms have derived from the "Green Revolution," it seems fair to conclude that the large farms have enjoyed greater benefits. The difference between the large and small farms, especially in the adoption and use of the new seeds and other inputs, is greater in the relatively more backward districts than in the more progressive ones. Therefore, while the income disparity between the large and small farms in all areas is likely to widen, it will be more pronounced in the backward districts.

ARE THE INCOME DISPARITIES BETWEEN THE LARGE AND SMALL FARMS AND BETWEEN THE PROGRESSIVE AND BACKWARD AREAS WIDENING?

In testing the preceding hypothesis, it was demonstrated that the benefits from the "Green Revolution" have been distributed unequally between the large and small farms in the Indus Basin of

Pakistan. Also, it was established that these benefits have been considerably greater in the progressive districts of the Punjab and Sind. What may perhaps be more significant is that the disparity of income between the large and small farms is greater in the more backward areas.

The unequal distribution of benefits is not the only source of widening interfarm and interregional income disparity. The fact that the income derived from the sale of inputs and outputs (other than crop output) tends to increase with the size of farm in most districts of the Punjab and Sind is another factor that will make the disparity even greater. However, it should be noted that the interfarm income disparity in the more backward districts would be reduced somewhat by the fact that in these districts, the income from other sources on small farms is quite high in relation to the income from crop output.

Combined with a skewed distribution of farm holdings, especially in Sind, the concentration of agricultural income will increase in the hands of a small number of farm owners. It also means that the relatively more progressive areas will continue to develop more rapidly in comparison with the more backward areas. The bimodal process of development is likely to be strengthened even more.

DO THE NEW WHEAT AND RICE SEEDS GENERATE MORE OR LESS EMPLOYMENT FOR HUMAN LABOR?

Several researchers asserted that the new wheat and rice seeds tend to generate more employment on the farms for both family and hired labor.[7] Of course, as yet there is very little evidence that such is the case in Pakistan. In this study, the index of employment is the number of man-days of family and hired labor per acre on a crop.

The data on the employment effects of the new wheat and rice seeds (Tables 13.3 and 13.4) lend no support to the hypothesis that they absorb more labor. For Mexi-Pak Wheat in the Punjab, it is quite clear that the employment of labor is less than on local varieties. In fact, the employment of family labor is significantly lower on Mexi-Pak wheat as compared with the local seeds. In Sind, on the other hand, more labor per acre is used on Mexi-Pak and a significantly higher level of employment is noted for hired labor.

In the Punjab, except for Jhelum, where there is a preponderance of family labor, more hired labor is used on both local and Mexi-Pak varieties of wheat. In general, more labor (both family and hired) is used on local varieties, except in Rahimyar Khan, where more hired labor is used on Mexi-Pak. The employment of family labor on both local and Mexi-Pak varieties is adversely

TABLE 13.3

Use of Human Labor per Acre, Wheat
(man-days per acre)

District/	Family Labor		Hired Labor		Total Labor	
Farm Size	Local	Mexi-Pak	Local	Mexi-Pak	Local	Mexi-Pak
Jhelum	77.1	6.0	4.0	26.0	81.1	32.0
< 12.50	107.8		3.4		111.2	
12.50-25.00	87.6		6.7		94.3	
25.00-50.00	67.6		2.4		70.0	
> 50.00		6.0		26.0		32.0
Gujranwala	6.2	4.4	11.9	8.3	18.1	12.7
< 12.50	12.0	9.8	6.0	4.3	18.0	14.1
12.50-25.00	5.8	6.6	4.2	9.1	10.0	15.7
25.00-50.00	8.6	4.3	14.3	5.4	22.9	9.7
> 50.00	4.8	2.6	9.3	4.7	14.1	7.3
Sahiwal	3.8	1.8	6.5	3.3	10.3	5.1
< 12.50	5.2	5.1	8.0	8.9	13.2	14.0
12.50-25.00	4.8	3.4	5.8	4.6	10.6	8.0
25.00-50.00	3.9	1.8	6.4	3.2	10.3	5.0
> 50.00	2.5	1.0	4.7	2.0	7.2	3.0
Lyallpur		3.5		6.7		10.2
< 12.50		9.6		5.1		14.7
12.50-25.00		6.3		10.5		16.8
25.00-50.00		4.4		5.7		10.1
> 50.00		2.0		3.4		5.4
Rahimyar Khan	16.8	11.5	15.6	23.6	32.4	35.1
< 12.50	19.3	17.7		16.7	19.3	34.4
12.50-25.00	20.2	13.7	3.0	10.8	23.2	24.5
25.00-50.00	15.2	10.1	10.4	10.0	25.6	20.1
> 50.00	10.0	10.8	11.0	21.8	21.0	32.6
Punjab	23.8	3.8	9.6	7.5	33.4	11.3
Jacobabad	7.1	6.3	1.1	2.3	8.2	8.6
< 12.50	16.5				16.5	
12.50-25.00	11.2	15.6			11.2	15.6
25.00-50.00	5.3	11.1			5.3	11.1
> 50.00	3.8	3.4	0.7	1.4	4.5	4.8
Larkana	6.2				6.2	
< 12.50	17.8				17.8	
12.50-25.00	11.2				11.2	
25.00-50.00	5.4				5.4	
> 50.00	3.3				3.3	
Nawabshah	6.6	6.1			6.6	6.1
< 12.50	17.9	27.2			17.9	27.2
12.50-25.00	13.1	6.6			13.1	6.6
25.00-50.00	4.9	5.7			4.9	5.7
> 50.00	3.1	2.8			3.1	2.8
Hyderabad		8.4		12.3		20.7
< 12.50		17.7		3.4		21.1
12.50-25.00		14.4		5.2		19.6
25.00-50.00		8.8		7.2		16.0
> 50.00		5.6		7.0		12.6
Sind	6.4	7.4	0.9	9.3	7.3	16.7

Note: Blank spaces indicate data not applicable.

Source: Compiled by the author.

TABLE 13.4

Use of Human Labor per Acre, Rice

District/ Farm Size	Family Labor			Hired Labor			Total Labor		
	Local	Basmati	IRRI	Local	Basmati	IRRI	Local	Basmati	IRRI
Jhelum									
< 12.50									
12.50-25.00									
25.00-50.00									
> 50.00									
Gujranwala	7.1	6.5	5.8	11.0	12.0	12.5	18.1	18.5	18.3
< 12.50	12.0	11.3	13.0	4.0	4.9	4.9	16.0	16.2	17.9
12.50-25.00	28.0	7.0	10.7	6.0	2.8	4.0	34.0	9.8	14.7
25.00-50.00	7.5	5.7	5.2	10.2	8.6	8.1	17.7	14.3	13.3
> 50.00	5.4	5.8	4.0	8.3	9.6	6.6	13.7	15.4	10.6
Sahiwal	3.4	3.0	3.1	4.3	4.9	5.4	7.7	7.9	8.5
< 12.50	7.0	7.6	3.0	9.0	6.7	4.0	16.0	14.3	7.0
12.50-25.00	4.7	5.2	4.9	2.3	7.7	8.9	7.0	12.9	13.8
25.00-50.00	3.5	3.1	3.8	2.2	4.9	7.1	5.7	8.0	10.9
> 50.00	2.0	1.9	2.1	1.3	3.2	3.3	3.3	5.1	5.4
Lyallpur	13.4	10.5		35.4	19.7		48.8	30.2	
< 12.50	100.0	17.3			6.7		100.0	24.0	
12.50-25.00	18.9	19.0		18.9	49.0		37.8	68.0	
25.00-50.00	10.7	15.6		18.7	23.9		29.4	39.5	
> 50.00		7.0			11.3		11.3	18.3	
Rahimyar Khan	21.0	15.0	28.0	3.9	10.0		24.9	25.0	28.0
< 12.50									
12.50-25.00	30.0		28.0				30.0		28.0
25.00-50.00	19.2	15.0		4.2	10.0		23.4	25.0	
> 50.0	21.0			4.0			25.0		
Punjab	7.5	4.6	5.5	10.6	9.2	12.2	18.1	13.8	17.7
Jacobabad	4.3		3.6	3.2		2.3	7.5		5.9
< 12.50	19.1		60.0				19.1		60.0
12.50-25.00	8.5		6.4				8.5		6.4
25.00-50.00	4.0		6.3				4.0		6.3
> 50.00	1.4		1.8	1.5		1.5	2.9		3.3
Larkana	14.7	26.5	5.1				14.7	26.5	5.1
< 12.50		61.0	26.8					61.0	26.8
12.50-25.00			9.3						9.3
25.00-50.00	7.5	30.3	3.5				7.5	30.3	3.5
> 50.00	18.0	16.2	2.0				18.0	16.2	2.0
Nawabshah									
< 12.50									
12.50-25.00									
25.00-50.00									
> 50.00									
Hyderabad	15.3		10.2	4.6		13.4	19.9		23.6
< 12.50	21.3		21.7	5.8		3.3	27.1		25.0
12.50-25.00	18.5		17.2	5.5		2.9	24.0		20.1
25.00-50.00	16.3		6.6	1.7		9.8	18.0		16.4
> 50.00	13.7		7.4	4.7		9.1	18.4		16.5
Sind	4.9	26.5	6.3	3.4		7.9	8.3	26.5	14.2

Note: Blank spaces indicate data not applicable.

Source: Compiled by the author.

affected by increased farm size: large farms tend to employ less labor, regardless of the variety of wheat in almost every district. Except in Rahimyar Khan, where it rises on Mexi-Pak, employment of hired labor is reduced with increased farm size for both local and Mexi-Pak wheat. The same trend is observed for the total labor per acre in every sample district of the Punjab.

In Sind there is no hired labor in Larkana and Nawabshah, and in Hyderabad more hired labor than family labor is used on Mexi-Pak wheat. In Jacobabad, however, less hired labor is used on both local and Mexi-Pak wheat. In general, more family labor is used in Sind, especially on local wheat. The use of labor, both family and hired, is adversely affected in every district as farm size increases: large farms use less labor per acre. In Hyderabad, however, hired labor for Mexi-Pak wheat increases with farm size.

The fact that in the Punjab less employment is provided by the new wheat seeds can be explained by the greater use of farm machinery. Since in Sind the use of farm machinery is far more limited on Mexi-Pak and local wheat, the adverse effect of the new seeds on employment is much smaller. In both provinces the use of labor declines on the large farms, which can best be explained by the fact that the large farms use more farm machinery per acre on the new wheat seeds. In general, large farms hire more labor for the Mexi-Pak seeds (as well as for local varieties) in the more progressive districts of the Punjab.

In the Punjab less labor is used on IRRI rice than on local rice, though more hired labor is used on the former. In Sind more labor, both family and hired, is used on IRRI rice than on local rice. Significantly higher use of hired labor is noted on the IRRI varieties in Sind.

In every Punjab district reporting rice, more hired labor is used than family labor, especially on IRRI rice. Family labor on both local and IRRI rice is used less on the large farms in every district. However, the use of hired labor, although it declines in Lyallpur and Rahimyar Khan as farm size increases, is greater on the large farms. The fact that more labor is hired on the large farms can be explained by the greater size of these farms, the stronger financial position of the owners, and the more limited use of family labor. In cases where the large farms employ less hired labor, there is a greater dependence on farm machinery. It must also be noted that in every district, the total use of labor, while slightly higher on IRRI rice, is less on large farms than small farms.

In Sind there is more use of family and hired labor (except in Hyderabad, where less hired labor is used) on local varieties compared with the new rice seeds. In every district except Hyderabad, where the large farms employ more hired labor for IRRI rice, the use of both family and hired labor is less on large farms than small farms. Only in Hyderabad is more labor used on IRRI rice than on local varieties.

The upshot of this analysis is that the new seeds of wheat and rice require less labor per acre than do the local seeds. This can best be explained by the substitution of machines for human labor on the large farms. This phenomenon is obviously more pronounced in the more progressive areas. It must, however, be noted that a greater amount of hired labor is used for the new seeds. The degree of displacement of labor by machines on the large farms is greater in the Punjab, a fact reflected in the data on the use of farm machinery for the new seeds.

HYPOTHESES GENERALIZED

The data generated by the field survey and their analysis in this chapter suggest several relevant and interesting generalizations on the likely consequences of the "Green Revolution" for other underdeveloped countries. For analytical convenience, these generalizations can be split into two groups. The first set emerges from the nature of the new technology. The second results from the economic and social consequences of the adoption process.

From a purely technical viewpoint, the "Green Revolution" is another name of the new biological-hydrological-chemical technology. It is significant to note that this technology is based on what may be called the complementarity principle: to realize the high-yield potential of the new seeds, they must be given more (or at least assured) water and chemical fertilizers. In areas and on farms where there is no assured supply of water and increased amounts of fertilizer are not used, the new seeds do not produce more per unit of land than do their local counterparts. This has been amply demonstrated in controlled experimental studies and in this study. The differences in the quantity of fertilizer and supplemental water used reflect the interregional and interfarm disparities in the output per unit of land and labor. The constraint on the use and profitability of the new seeds is seen most clearly in the rain-fed areas (as in Pakistan) and in areas where frequent flooding occurs (as in some tropical countries).

The adoption and use of the new seeds is also constrained by economic and social factors, with consequent growth and income disparities among regions and farms. In general, the new technology has been accepted and used more readily in regions where commercialized agriculture already exists, and consequently it benefits these regions more than others. Further, within each region it affects the large farms more favorably. In fact, the new technology tends to strengthen the existing commercialized farming, often at the expense of peasant (mainly subsistence) farming. Where peasant farming is predominant, the new technology is not suitable, through either the lack of one or more complementary inputs or government policies discriminating in favor of the large farms.

The fact that often the more prosperous farmers (who are also the owners of large farms) are the ones who innovate--the others can at best follow them--is explained by their stronger financial position and their superior knowledge of, and greater access to, credit and new inputs. They have achieved these advantages through their social and political power. The small farmers and peasants are caught in a double squeeze. In many countries where the government policies of subsidizing agriculture favor the large landowners, the effect of increasing return on the land provides a strong incentive to acquire land from small farmers who cannot compete in obtaining the necessary inputs. The second source of pressure is in the increasing tendency of the large landowners to use labor-saving innovations that in general decrease the demand for farm labor. The inexorable pressure of population on land, resulting in the fragmentation of holdings and the increased number of landless workers, makes the polarization of classes in these countries more imminent.

Yet another dimension of the cost of the "Green Revolution" is the fact that the innovative farmers have enjoyed substantial subsidies on inputs and price supports on outputs. Input subsidies obviously entail the transfer of resources from alternative uses with high opportunity cost. The price support system adds to the problem of allocative efficiency. Integrating the principles of domestic resource allocation and comparative advantage in international trade, the price of additional food grains that the society is made to pay in several underdeveloped countries may have to be reexamined. It is suspected by many that under a system of subsidies, while the private benefits accruing to the innovative farmers are increasing, the social costs may in fact be higher than the social benefits from the adoption and use of the new technology.

NOTES

1. See, for instance, H. Kaneda, "Economic Implications of the 'Green Revolution' and the Strategy of Agricultural Development in West Pakistan," Pakistan Development Review 9, no. 2 (Summer 1969): 111-43; B. F. Johnston and J. Cownie, "The Seed-Fertilizer Revolution and Labor Force Absorption," American Economic Review 59, no. 4 (September 1969): 569-82; J. Cownie, B. F. Johnston, and B. Duff, "The Quantitative Impact of the Seed-Fertilizer

Revolution in West Pakistan: An Exploratory Study," Food Research Institute Studies in Agricultural Economics, Trade and Development 9, no. 1 (1970): 57-95; W. P. Falcon, "The Green Revolution: Generations of Problems," American Journal of Agricultural Economics 52, no. 5 (December 1970): 698-710; L. Nulty, The Green Revolution in West Pakistan: Implications of Technological Change (New York: Praeger, 1972); M. H. Khan, "'Green Revolution' or 'Technocratic Euphoria': Some Problems of Rapid Agricultural Change in Asia," Economia Internazionale 26, no. 2 (February 1973): 1-17.

2. See Falcon, op. cit., p. 707. Since the data collected for the 1972 Census of Agriculture had not been made public as this book was going to press, the figures of the 1960 Census on the distribution of farm holdings were the only ones available. It is commonly believed in Pakistan that, despite the Land Reform Law of 1972 instituted by the present leadership, the distribution of holdings has not changed significantly since the land reforms of 1959-60, during the Ayub regime.

3. Esso Pakistan Fertilizer Co., Ltd., "Pakistan Nitrogen Demand Forecast Study, 1974-75 to 1984-85" (Karachi: Esso, October 1974). (Mimeographed.)

4. Falcon, op. cit., p. 706.

5. These figures were furnished by the Agriculture and Food Section, Pakistan Planning Commission, Islamabad.

6. Although the issue of subsidy on canal water is still heatedly debated in Pakistan, there is a well-documented study on the extent of subsidy on the use of canal water: M. S. Chaudhry, "Economic Impacts of Wheat Yield Increases in West Pakistan" (unpublished Ph.D. dissertation, Washington State University, 1971). The canal water rates for various crops in the Punjab in no way reflect the marginal value of water to the farmer. The flat water-rate system in Sind, introduced in 1972, seems to be even more regressive. In both systems the major beneficiaries are often the large farms, which, compared with the small farms, devote a higher percentage of the acreage to cash and high-value crops.

7. See, for instance, Johnston and Cownie, op. cit.; and Falcon, op. cit.

CHAPTER

14

CONCLUSION

This book has examined the adoption and use of the new seeds of wheat and rice and its effect on income and employment on different farm sizes in the Indus Basin of Pakistan. The hypotheses tested are the following:

1. Is the participation in the adoption and use of the new inputs by large and small farms different?
2. Are the new wheat and rice seeds more profitable than the local seeds, and is there any difference in their profitability on large and small farms?
3. Are the benefits from the "Green Revolution" distributed unequally between large and small farms?
4. Are the income disparities between large and small farms and between progressive and backward areas widening?
5. Do the new wheat and rice seeds generate more employment for farm labor?

The answers to these questions yield interesting conclusions. Some of them reinforce the accepted conventional wisdom, while others help demolish popular myths. The major findings of this study can now be summarized.

1. The new wheat and rice seeds are more profitable than the local seeds.
2. Even though the new seeds may not necessarily be more profitable to large farms than small farms, the benefits from the "Green Revolution" to large farms tend to be greater.
3. The income disparities between large and small farms are likely to widen, more so in the relatively more backward agricultural regions.

4. The interregional income disparities are likely to increase.

5. Looking at the yield-per-acre figures in Table 4.3, one would be led to believe that small farms are the less efficient units. However, if a more appropriate index of economic efficiency is used (average value of farm output per acre divided by average variable cost per acre), the small farms are either more efficient than the large farms or at least equally efficient. These conclusions are strongly supported by the data given in Table A. 17.

6. The new technology, used in the production of Mexi-Pak wheat and IRRI rice, is by no means neutral to scale.

7. The new wheat and rice seeds provide less employment on the farm than the local seeds, which can be explained (but only in part) by the displacement of human labor by machines.

8. Considering the levels of inputs used on many farms (especially chemical fertilizer), the high yield of the new seeds may be decreasing, unless the current farm management practices are changed radically or the inherent vigor of the new seeds is maintained by further breeding.

If these conclusions are correct, as the results of this study would indicate, they do not augur well for diffused agricultural growth in Pakistan. The divergence of benefits to large and small farms and the disparities between the progressive and backward districts in the Indus Basin of Pakistan must be halted by specific policy measures.[1] To achieve the dual goal of growth and equity, I believe that the policy-makers in Pakistan must adopt measures in at least the following areas.

THE LAND TENURE SYSTEM

The kinds of changes in the land tenure system in Pakistan that need to be emphasized are penalizing the absentee landlord; increasing the security to the tenant and, more important, providing incentives to him through direct assistance; and halting the fragmentation of holdings by maintaining a minimum size. The land reforms of 1972, which were instituted by the present government, seem to have been motivated primarily by the electoral pledge of the Pakistan People's Party in 1970. At best, these reforms have increased the power given to the tenant and have resulted in some redistribution of the appropriated farm land. However, the new owners or the sharecroppers have not been given the wherewithal in terms of credit and inputs. That the land reforms of 1972 have had little if any effect on the economic and political structure in the countryside is well reflected by the feudal character of the political leadership of the ruling party in the Punjab and Sind. Growth and its diffusion can be increased only if incentives and disincentives form an integral part of the land ownership system.

TAX ON AGRICULTURAL INCOMES

In Pakistan in the last two years, probably there is no other question that has been debated as heatedly, and to which the leadership has been as sensitive, as the introduction of tax on agricultural incomes. The fact that the present regime has not been able to do so indicates the increased (rather than decreased) power of the feudal elite in the political system. The incomes on large farms, which have increased mainly through the use of a subsidized package of technology, when compared with incomes in the nonagricultural sectors, reveal not only that an important source of "agricultural surplus" remains untapped for public investment but also that the taxation system is patently inequitable. In fact, the revenue from the land tax, which has been the only source of income to the government from the agricultural sector, has been declining in proportion to the national income: from 1.7 percent to 1.0 percent during 1960-70.[2] The increased incomes in agriculture, which remain untaxed, find their way into not-so-hidden consumption, thus contributing to an already high rate of inflation.

What otherwise would be an investable surplus, estimated to be in the range of Rs. 700 million to Rs. 800 million, remains mostly in the hands of those who have received increased incomes as a result of the subsidies given by the society. It should also be noted that at least in the last two years the terms of trade have moved in favor of the agricultural sector. It seems to be an opportune time to study more seriously the extent to which agricultural incomes could be taxed without adversely affecting the incentives for increased investment and production. The impression that I gathered in Pakistan is that agricultural incomes remain untaxed not because taxing them is economically unsound but it is politically inexpedient.

ESTABLISHMENT OF MARKETING BOARDS

Most agricultural products are channeled through unregulated markets on the village level. It would seem appropriate to establish regional marketing boards for at least the major grain (wheat and rice) and cash (cotton and sugarcane) crops. The primary functions of these boards should be to procure marketable surpluses at guaranteed and reasonable prices, and to provide credit and farm inputs against the produce that farmers deliver. These boards should be most active in performing these functions for the small farm operators, who have little access to institutional credit and farm inputs and who also are victims of the rapacious middlemen operating on the village or district level.

AGRICULTURAL EXTENSION SERVICE

The agricultural extension service is no different from any other bureaucratic agency of the government. It is the owners of large farms, and the areas where progress is more obvious, that use the services of field agents. Those who operate small farms often remain deprived of this service. It is, therefore, imperative to transform the extension service into a market-oriented organization that can help farmers, especially the operators of small farms, in planning and production.

FARM MECHANIZATION

The increasing use of imported farm machinery in Pakistan, which the large farms have found so profitable (thanks to the generous subsidies) and which has tended to displace human labor, needs serious reexamination. The use of machines has not made the large farms necessarily more efficient, and the benefits from the use of the new biological-hydrological-chemical technology do not depend on the increased use of farm machinery. Economic rationality would seem to suggest the development of indigenous technology suited to the resource endowment of the country.

RESEARCH ON THE NEW SEEDS

The two disturbing features of the new wheat and rice seeds are that the high yields of these seeds can be maintained only by selective breeding, and that they are more susceptible to pests and diseases than the local seeds. Also, since the higher-yield potential of the new seeds depends on the availability of water, the rain-fed regions (such as Jhelum) have been excluded from the benefits of increased incomes. Active research is needed in water management and in the development of new strains of wheat that can show a better performance than the local seeds.

The food crisis that Pakistan faces, on the heels of the "Green Revolution," can be averted by adopting enlightened institutional and pricing policies. Recent changes in the policies affecting the prices of inputs and outputs inspire some confidence. However, there has been little evidence of institutional changes, the prime examples of which are the land tenure system and tax on agricultural incomes.

The fact that agriculture in the Indus Basin, which irrigates an area of about 33 million acres, cannot feed a population of about 66 million is a sad reflection on the agricultural system of Pakistan. Farm efficiency is still very low, but there is scope for its improvement, especially on the small farms, if appropriate institutional changes take effect. Technology, however profitable it may be, cannot operate in a vacuum. It must be made profitable for the individual farmer and economic for the society. Farmers in Pakistan, even those who operate the small farms, have exhibited rational behavior in their allocation and production decisions. To make the process of change participatory and profitable for millions of farmers, the constraints they face must be removed by concerted action. Growth and equity without costs are impossible to achieve in most circumstances. The test of good leadership lies in its ability to reduce these costs.

A word researchers. Given the state of our knowledge (which at present is indeed limited) about resource allocation and production decisions on the farm level, it seems quite unfair to make large-scale generalizations. The areas of economic research that need to be explored more thoroughly in countries like Pakistan are related to the input-output relationships on the farm; the physical and financial constraints that the farmers face, especially on the small farms; and the alternative models for resource combination, with a view to increasing farm efficiency and profitability. Macro-level decisions must be based not on generalizations alone. For these decisions to be rational and optimum, economists must provide the policymakers with the micro-level inputs that go into determining the allocation and production decisions of the farmers.

NOTES

1. For an excellent exposition on public policy related to agricultural development, see K. Griffin, "Policy Options for Rural Development," Oxford Bulletin of Economics and Statistics 35, no. 4 (November 1973): 239-74.

2. Ibid., p. 248.

APPENDIX A

CANAL WATER COSTS IN THE PUNJAB AND SIND

In Table A.1, the water charge per acre by crop for each sample district of the Punjab is given. Knowing the area of each crop irrigated by the respondents and the water charge per acre for each crop, it has been possible to calculate the cost of canal water.

In Table A.2, the water rate per acre in Sind is given. In the flat-rate system, which was introduced in Sind in 1972-73, the provincial government has specified the "produce index unit" per acre for the various districts. A "produce index unit" is defined as the ratio of the gross value of the produce and the total matured area multiplied by the ratio for the district. Thus the water rate in each district has been determined on the basis of the "produce index unit" in all districts. In the table the average for each farm category reflects the water charges for the total area and per acre.

TABLE A.1

Charges for Irrigation Water per Acre in the Punjab
(rupees)

Crop	Sahiwal	Gujranwala and Lyallpur	Rahimyar Khan
Wheat	10.40	10.40	10.40
Maize	9.60	9.60	8.00
Gram	8.00	8.00	8.00
Oilseeds	12.00	12.00	12.00
Fodder	6.40	6.40	6.40
Fruits, vegetables	20.80	20.80	16.80
Other	10.40	10.40	10.40
Rice	16.00	16.00	14.40
Cotton	16.80	16.00	16.80
Sugarcane	32.80	32.80	31.20

Source: Government of West Pakistan, Irrigation and Power Department, *Gazette Notification*, Extraordinary issue, *Revised Schedule of Occupiers Rates 1969*, September 29, 1969.

TABLE A.2

Charges for Irrigation Water per Acre in Sind
(rupees)

District/ Farm Size	Average Cost of Canal Water	Average Size of Holding (acres)	Average Cost of Canal Water
Jacobabad			
Under 12.50	68.48	9.3	7.36
12.50-25.00	137.00	19.1	7.17
25.00-50.00	262.37	35.3	7.43
Over 50.00	612.13	77.0	7.50
Larkana			
Under 12.50	90.81	8.0	11.35
12.50-25.00	206.30	19.3	10.69
25.00-50.00	393.77	36.6	10.76
Over 50.00	785.84	65.0	12.09
Nawabshah			
Under 12.50	88.28	8.0	11.04
12.50-25.00	195.59	17.3	11.31
25.00-50.00	417.82	36.1	11.57
Over 50.00	872.38	75.0	11.63
Hyderabad			
Under 12.50	90.96	9.7	9.38
12.50-25.00	173.33	19.3	8.98
25.00-50.00	312.37	36.0	8.68
Over 50.00	756.18	84.0	9.0

Notes: The calculations of average cost of canal water for the different farm sizes are based on the water rates charged by the Government of Sind in 1972-73.

The figures for the average size of holding are taken from Table 3.1.

Source: Government of Sind, Board of Revenue, Flat Rates per Produce Index Unit of Land Revenue and Water Rate (Hyderabad: the Board, September 1973).

APPENDIX B

SURVEY VILLAGES AND DISTRIBUTION OF RESPONDENTS

This appendix gives the names and numbers of villages where the interviews were conducted (Table A.3) and the distribution of respondents in the sample districts of the Punjab and Sind (Table A.4).

TABLE A.3

Villages in the Sample Districts

District/ Province	Villages	Number of Villages
Jhelum	Dharyala; Saroba; Thir Chak; Taniala	4
Gujranwala	Kotli Mughalan; Kotli Diyanat Rai; Khotra; Jajoke; Kingaryai; Sadhoke; Chak Santa; Khanki; Kot Nawan; Thatti Anokh Singh; Khurram Lodhi; Hardo Ratta; Muradian; Rakh Pindi Jalol	14
Sahiwal	Chak 33/S.P.; Sadhu Maleke; Chak Kot Bakhsha; Chak 27/K.B.; Pakhian; Chak 44/E.B.; Chak 361/E.B.; Chak 29-30/2-L; Chak 33/G.D.; Lehri Zereen; Chak 6/1-AL; Chak 54/12-L; Chak 111/9-L; Chak 6/14-L; Chak 34/12-L; Chak 137/9-AL; Chak 60-61/G.D.; Khangranwala; Amli Moti; Chak Chishti Urf Ganj Bux; Jodha; Shah Yakka; Mali Mahar	23
Lyallpur	Baharawala; Nawade; Chak 668/G.B.; Kot Guru Sahib; Nakreri; Sham Fatuwal; Plot 56; Manka; Jhok Rahman; Kot Jahan Khan; Chak Lishari; Galfabad; Kukkar; Sarfatu; Kamalpur Kalan; Jasoana; Chakku; Dauana; Lokha Kalan	19
Rahimyar Khan	Chak 87/N.P.; Chak 115/N.P.; Garhi Khair Mohammad Jhak; Sanjarpur Nao; Chak 236/P; Chak 102/N.P.; Amir Pur; Chak 47/Abbasya; Tarinda Gur Gaij; Bhapla Kacha	10
Punjab		70
Jacobabad	Udi; Samejo; Hazaro; Wassayo; Pako Kashmore; Pako Khoski	6
Larkana	Daro Dodaiko; Potho Ibrahim; Seer Drakhan; Zangeja; Khaliq Dino Dakhan; Nasirabad	6
Nawabshah	Nasrat; Nasrat; Shuja Mohammad; Tharo Unar; Bakheri No. 1; New Gachero; Kandhari	7
Hyderabad	Andhalo; Dha Machani; Abri; Khar Yoon; Noochani; Kari Mohammed Ali; Dethki; Jiandal Kote; Kamaro Sharif; Gad; Kuhrai	11
Sind		30

Source: Compiled by the author.

TABLE B.4

Distribution of Respondents in the Sample Districts

District and Province	Number of Respondents					
	Under 12.50 Acres	12.50–25.00 Acres	25.00–50.00 Acres	Over 50.00 Acres	Landless Workers	Total Number
Jhelum	8	7	3	2	8	28
Gujranwala	28	24	28	12	28	120
Sahiwal	46	46	46	46	46	230
Lyallpur	38	38	38	34	38	186
Rahimyar Khan	20	20	20	14	20	94
Punjab	140	135	135	108	140	658
Jacobabad	11	12	12	12	12	59
Larkana	11	12	12	12	12	59
Nawabshah	14	14	14	14	14	70
Hyderabad	22	22	20	20	22	106
Sind	58	60	58	58	60	294
Punjab and Sind	198	195	193	166	200	952

Source: Compiled by the author.

APPENDIX C

QUESTIONNAIRE

Interviewer__________________
Date________________________

1. Name of Farmer:____
2. Village:__________ 3. Tehsil:________ 4. District:__________
5. Total Area of Holding: ________ Acres
6. Area Rented Out: __________ Acres
7. Area Rented In: __________ Acres
8. Area Self-Cultivated: ____________ Acres
9. Is the Self-Cultivated Area Fragmented? Yes____ No ____
10. Number of Fragments: 1 2 3 4 5 6 7 8 9 ______
11. Holding Held: Individually ____ Jointly _____
12. Ownership: Owner _____ Tenant _____ Lessee _____
13. If Tenant, Crop Share: ______ Percent
14. If Lessee, Rent per Acre: ________ Rupees
15. Family Size: ______ Persons
    Adults _____ Minors _____
    Male ___ Female ___ Male ___ Female ___
16. Family Members Working: _____ Persons
    On the Farm _____ Off the Farm _____
    Full Time ___ Part-Time ___ Full-Time ___ Part-Time ___
17. Permanent Hired Workers: _____ Persons
18. Work Animals: _______
    Bullocks ___ Other ____
    Work ___ Non-Work ___ Work ___ Non-Work ___
19. Milch Animals: _____
    Cows ___ Buffaloes _____
    Milch ___ Non-Milch ___ Milch ___ Non-Milch ____
    Other
    Milch ___ Non-Milch ___

20. Cropping Pattern:

Crop	Area Sown (acres)	Production (maunds)	Disposal of Output: Quantity for Seed (maunds)	Disposal of Output: Quantity for Consumption (maunds)	Disposal of Output: Quantity for Market (maunds)	Price of Output (Rs. per maund)
Rabi 1972-73						
Wheat						
Local						
Mexi-Pak						
Maize						
Gram						
Pulses						
Oilseeds						
Fodder						
Vegetables						
Fruits						
Other						
Kharif 1973						
Rice						
Local						
IRRI						
Basmati						
Cotton						
Local						
Improved						
Sugarcane						
Maize						
Local						
Improved						
Pulses						
Oilseeds						
Fodder						
Vegetables						
Fruits						
Other						

20. Cropping Pattern (continued):

Crop	Seed			Water (Canal)		
	Amount (maunds)	Seed per Acre (seers)	Total Price (Rs.)	Number of Irri- gations	Area Irri- gated (acres)	Water Rate per Acre (Rs.)
Rabi 1972-73						
Wheat						
Local						
Mexi-Pak						
Maize						
Gram						
Pulses						
Oilseeds						
Fodder						
Vegetables						
Fruits						
Other						
Kharif 1973						
Rice						
Local						
IRRI						
Basmati						
Cotton						
Local						
Improved						
Sugarcane						
Maize						
Local						
Improved						
Pulses						
Oilseeds						
Fodder						
Vegetables						
Fruits						
Other						

20. Cropping Pattern (continued):

Crop	Water					
	Tubewell			Surface Well		
	Number of Irrigations	Area Irrigated (acres)	Water Rate per Acre (Rs.)	Number of Irrigations	Area Irrigated (acres)	Price per Acre Irrigated (Rs.)
Rabi 1972-73						
Wheat						
Local						
Mexi-Pak						
Maize						
Gram						
Pulses						
Oilseeds						
Fodder						
Vegetables						
Fruits						
Other						
Kharif 1973						
Rice						
Local						
IRRI						
Basmati						
Cotton						
Local						
Improved						
Sugarcane						
Maize						
Local						
Improved						
Pulses						
Oilseeds						
Fodder						
Vegetables						
Fruits						
Other						

20. Cropping Pattern (continued):

Crop	Fertilizer					
	Organic (FYM)		Chemical			
			Nitrogen		Phosphorus	
	Quantity Used (cartload)	Rate per Cartload (Rs.)	Quantity Used (bags)	Rate per Bag (Rs.)	Quantity Used (bags)	Rate per Bag (Rs.)
Rabi 1972-73						
Wheat						
Local						
Mexi-Pak						
Maize						
Gram						
Pulses						
Oilseeds						
Fodder						
Vegetables						
Fruits						
Other						
Kharif 1973						
Rice						
Local						
IRRI						
Basmati						
Cotton						
Local						
Improved						
Sugarcane						
Maize						
Local						
Improved						
Pulses						
Oilseeds						
Fodder						
Vegetables						
Fruits						
Other						

20. Cropping Pattern (continued):

	Human Labor					
			Hired			
	Family		Permanent		Casual	
Crop	Man-Days	Rate per Man-Day (Rs.)	Man-Days	Rate per Man-Day (Rs.)	Man-Days	Rate per Man-Day (Rs.)
Rabi 1972-73						
Wheat						
Local						
Mexi-Pak						
Maize						
Gram						
Pulses						
Oilseeds						
Fodder						
Vegetables						
Fruits						
Other						
Kharif 1973						
Rice						
Local						
IRRI						
Basmati						
Cotton						
Local						
Improved						
Sugarcane						
Maize						
Local						
Improved						
Pulses						
Oilseeds						
Fodder						
Vegetables						
Fruits						
Other						

20. Cropping Pattern (continued):

	Animal Labor							
	Owned				Hired			
	Bullocks		Other		Bullocks		Other	
Crop	Area Operated (acres)	Rate per Acre (Rs.)	Area Operated (acres)	Rate per Acre (Rs.)	Area Operated (acres)	Rate per Acre (Rs.)	Area Operated (acres)	Rate per Acre (Rs.)
Rabi 1972-73								
Wheat								
Local								
Mexi-Pak								
Maize								
Gram								
Pulses								
Oilseeds								
Fodder								
Vegetables								
Fruits								
Other								
Kharif 1973								
Rice								
Local								
IRRI								
Basmati								
Cotton								
Local								
Improved								
Sugarcane								
Maize								
Local								
Improved								
Pulses								
Oilseeds								
Fodder								
Vegetables								
Fruits								
Other								

20. Cropping Pattern (continued):

	Farm Machinery							
	Tractors				Other			
	Owned		Hired		Owned		Hired	
Crop	Area Operated (acres)	Rate per Acre (Rs.)	Area Operated (acres)	Rate per Acre (Rs.)	Area Operated (acres)	Rate per Acre (Rs.)	Area Operated (acres)	Rate per Acre (Rs.)
Rabi 1972-73								
Wheat								
Local								
Mexi-Pak								
Maize								
Gram								
Pulses								
Oilseeds								
Fodder								
Vegetables								
Fruits								
Other								
Kharif 1973								
Rice								
Local								
IRRI								
Basmati								
Cotton								
Local								
Improved								
Sugarcane								
Maize								
Local								
Improved								
Pulses								
Oilseeds								
Fodder								
Vegetables								
Fruits								
Other								

20. Cropping Pattern (continued):

Crop	Marketing		Remarks
	Haulage Costs (Rs.)	Other Costs (Rs.)	
Rabi 1972-73			
Wheat			
Local			
Mexi-Pak			
Maize			
Gram			
Pulses			
Oilseeds			
Fodder			
Vegetables			
Fruits			
Other			
Kharif 1973			
Rice			
Local			
IRRI			
Basmati			
Cotton			
Local			
Improved			
Sugarcane			
Maize			
Local			
Improved			
Pulses			
Oilseeds			
Fodder			
Vegetables			
Fruits			
Other			

21. Tubewell Owned: Yes _____ No _____
    If Yes:
    Diesel ___ Electric ___
    Capacity of Delivery Pipe ___ HP ___
    Year of Installation _______
    Initial Cost of Installation _____ Rupees
    Operating Cost for 1972-73 _____ Rupees
    Labor _____ Spare Parts _____ Fuel _____ Other _____
22. Tractor Owned: Yes ___ No ___
    If Owned:
    Make ____________________ HP _______
    Year of Purchase _______
    Initial Cost _______ Rupees
    Operating Cost for 1972-73 _______ Rupees
    Labor _____ Spare Parts _____ Fuel _____ Other ______

23. Tractor-Drawn Implements:

Implement	Number	Type	Cost per Piece (Rs.)
Plough			
Cultivator/tiller			
Harrow			
Planter			
Trailer			
Other			

24. Animal-Drawn Implements:

Implement	Number	Source of Supply			Cost per Piece (Rs.)
		Village	Mandi	Other	
Iron plough					
Cultivator/tiller					
Drill					
Fodder cutter					
Cane cutter					
Harrow					
Other					

25. Fertilizers:

Type of Fertilizer	Quantity Used (bags/ cartloads)	Rate per bag/ cartload (Rs.)	Source of Purchase		
			Village	Mandi	Other
Non-chemical					
FYM					
Other					
Chemical					
Nitrogen					
Phosphorus					
Other					

26. Pesticides:

Type of Pesticide	Acres Treated	Cost per Acre (Rs.)	Source of Supply		
			Village	Mandi	Other
1					
2					
3					
4					

27. Markets:

Name of Market	Suppler of Input	Sale of Output	Source of Credit	Distance from Farm (miles)	Nature of Road	
					Katcha	Pucca
1						
2						
3						
4						

28. Debt Situation:

Source	Amount of Loan (Rs.)	Month/ Year Taken	Pur- pose	Interest Rate (%)	First Install- ment Due	Last Install- ment Due
Public						
Agric. Dev. Bank						
Cooperative						
Taccavi						
Private						
Commission agent						
Relatives, friends						
Moneylenders						
Other						

29. Carry-Over Stocks of Seed and Fertilizer:

Kind of Seed	Quantity (maunds)	Kind of Fertilizer	Quantity (nutrient lbs.)
1			
2			
3			
4			

30. Off-Farm Earnings:

a. Sale of Labor Service: Yes _____ No _____

If Yes:

Number of Man-Days Worked _____

Rate per Man-Day _______ Rupees

b. Sale of Animal Labor: Yes _____ No _____

If Yes:

Number of Days/Acres Worked _____

Rate per Day/Acre ______ Rupees

c. Sale of Tractor/Machine Service: Yes ____ No ____

If Yes:

Type of Machine	Hours Rented	Rate per Hour (Rs.)
1		
2		
3		
4		

d. Sale of Tubewell Water: Yes _____ No _____
If Yes:
Number of Hours Sold ______
Rate per Hour _______ Rupees

e. Sale of Transport Service: Yes _____ No _____
If Yes:

Type of Transport	Mileage	Rate per Mile (Rs.)
1		
2		
3		
4		

f. Other Business: Yes _____ No _____
If Yes:

Type of Business	Income per Day/ Month	Number of Days/ Months	Total Income (Rs.)
1			
2			
3			

31. When were the following inputs first used, and what were the sources of their supply?

Input	Year First Used	Source of Supply		
		Village	Mandi	Other
Chemical fertilizer				
Nitrogen				
Phosphorus				
Other				
New seeds				
Wheat				
Rice				
Maize				
Pesticides				
Farm machinery				
Tractor				
Other				
Tubewell water				

32. Landless Workers:

	Employed				
	Agriculture		Nonagriculture		
Type of Job	Number of Man-Days	Income (Rs.)	Number of Man-Days	Income (Rs.)	Days Unemployed
1					
2					
3					
4					

TABLE A.5

Adoption and Use of New Inputs
(Jhelum)

Input	% Never Used	% Used Once	Percentage Farmers Using: 1972-73	2 Yrs.	4 Yrs.	6 Yrs.	8 Yrs.	Over 8 Yrs.	% Never Used	% Used Once	Percentage Farmers Using: 1972-73	2 Yrs.	4 Yrs.	6 Yrs.	8 Yrs.	Over 8 Yrs.
	Under 12.50 Acres								12.50-25.00 Acres							
Fertilizer																
Nitrogenous	87.5	12.5						12.5	71.0	29.0			14.0	14.0		
Phosphatic	100.0								100.0							
Other	100.0								100.0							
New seeds																
Wheat (Mexi-Pak)	87.5	12.5			12.5				100.0							
Rice (IRRI)	100.0								100.0							
Pesticides	100.0								100.0							
Farm machinery																
Tractor	100.0								71.0	29.0	14.0	14.0				
Other	75.0	25.0			12.5		12.5		100.0							
Tubewell water	100.0								100.0							
	25.00-50.00 Acres								Over 50.00 Acres							
Fertilizer																
Nitrogenous	100.0								50.0	50.0	50.0					
Phosphatic	100.0								100.0							
Other	33.4	66.7						66.7	50.0	50.0					50.0	
New seeds																
Wheat (Mexi-Pak)	100.0								50.0	50.0	50.0					
Rice (IRRI)	100.0								100.0							
Pesticides	100.0								100.0							
Farm machinery																
Tractor	100.0									100.0			50.0		50.0	
Other	66.7	33.3		33.3						100.0			50.0		50.0	
Tubewell water	100.0								100.0							

(continued)

TABLE A.5 (continued)
(Gujranwala)

Input	% Never Used	% Used Once	Percentage Farmers Using 1972-73	2 Yrs.	4 Yrs.	6 Yrs.	8 Yrs.	Over 8 Yrs.	% Never Used	% Used Once	Percentage Farmers Using 1972-73	2 Yrs.	4 Yrs.	6 Yrs.	8 Yrs.	Over 8 Yrs.
	Under 12.50								12.50-25.00 Acres							
Fertilizer																
Nitrogenous	3.6	96.4			60.7	35.7				100.0				4.2	8.3	87.5
Phosphatic	100.0								33.3	66.7				4.2	8.3	54.2
Other	100.0								100.0							
New seeds																
Wheat (Mexi-Pak)		100.0			46.4	53.6				100.0			25.0	8.3	37.5	29.2
Rice (IRRI)		100.0		3.6	96.4					100.0			25.0	37.5	33.3	4.2
Pesticides	14.3	85.7			85.7				41.7	58.3			4.2	8.3	20.8	25.0
Farm machinery																
Tractor	75.0	25.0			21.4	4.6			100.0							
Other	100.0								100.0							
Tubewell water	67.9	32.1			21.4	10.7			33.3	66.7				25.0	20.8	20.8
	25.00-50.00 Acres								Over 50.00 Acres							
Fertilizer																
Nitrogenous		100.0				3.6	21.4	75.0		100.0				50.0	50.0	
Phosphatic	64.3	35.7		3.6	7.1	10.7	3.6	10.7	66.7	33.3	8.3		25.0			
Other	100.0								100.0							
New seeds																
Wheat (Mexi-Pak)	3.6	96.4			17.9	28.6	35.7	14.3		100.0			41.7	58.3		
Rice (IRRI)		100.0		7.1	64.3	25.0	3.6			100.0			41.7	58.3		
Pesticides	89.3	10.7				3.6	3.6	3.6	50.0	50.0		16.7	33.3			
Farm machinery																
Tractor	82.2	17.9		3.6	3.6	3.6	7.1		8.3	91.7	8.3	8.3	75.0			
Other	100.0								100.0							
Tubewell water	28.6	71.4		3.6	14.3	21.4	25.0	7.1		100.0		8.3	75.0	16.7		

(Sahiwal)

	Under 12.50 Acres								12.50-25.00 Acres							
Fertilizer																
Nitrogenous		100.0		6.5	21.7	37.0	17.4	17.4		100.0			23.8	35.8	31.0	9.5
Phosphatic	58.7	41.3	8.7	19.6	6.5	4.4	2.2		33.3	66.7	11.9	21.4	28.6	2.4	2.4	
Other	100.0								100.0							
New seeds																
Wheat (Mexi-Pak)	4.4	95.7		13.0	58.7	21.7	2.1			100.0			66.7	33.3		
Rice (IRRI)	45.7	54.3		13.0	37.0	4.4			28.6	71.4		9.5	40.5	21.4		
Pesticides	97.8	2.2		2.2					92.9	7.1		4.8	2.4			
Farm machinery																
Tractor	93.5	6.5		2.2	4.4				100.0							
Other	100.0															
Tubewell water	78.3	21.7			21.2	21.2	4.4	13.0	45.2	54.8		4.8	21.4	9.5	9.5	9.5
	25.00-50.00 Acres								Over 50.00 Acres							
Fertilizer																
Nitrogenous		100.0			15.2	26.1	28.3	30.4		100.0			6.7	20.0	33.3	40.0
Phosphatic	10.9	89.1	4.4	30.4	39.1	8.7	6.5		8.9	91.1	4.4	15.6	53.3	13.3	4.4	
Other	100.0								100.0							
New seeds																
Wheat (Mexi-Pak)		100.0			56.5	41.3	2.2			100.0			60.0	37.8	2.2	2.2
Rice (IRRI)	23.9	76.1		13.0	43.5	17.4		2.2	20.0	80.0	11.1	51.1	17.8			
Pesticides	65.2	34.8	10.9	21.7			2.2		40.0	60.0	6.7	33.3	11.1		8.9	
Farm machinery																
Tractor	76.1	23.9	2.2	10.9	10.9				4.4	96.0	4.4	35.6	48.9	4.4	2.2	
Other	100.0								100.0							
Tubewell water	8.7	91.3		4.4	32.6	21.7	8.7	23.9	2.2	97.8			37.8	28.9	13.3	17.8

(continued)

TABLE A.5 (continued)
(Lyallpur)

Input	% Never Used	% Used Once	Percentage Farmers Using 1972-73	2 Yrs.	4 Yrs.	6 Yrs.	8 Yrs.	Over 8 Yrs.	% Never Used	% Used Once	Percentage Farmers Using 1972-73	2 Yrs.	4 Yrs.	6 Yrs.	8 Yrs.	Over 8 Yrs.
	Under 12.50 Acres								12.50-25.00 Acres							
Fertilizer																
Nitrogenous		100.0		5.3	57.9	36.8				100.0			13.2	73.7	13.2	
Phosphatic	94.7	5.3				5.3			79.0	21.1			13.2	5.3	2.6	
Other	100.0								100.0							
New seeds																
Wheat (Mexi-Pak)		100.0		5.3	57.9	36.8				100.0			13.2	86.8		
Rice (IRRI)	100.0								81.6	18.4			15.8	2.6		
Pesticides	100.0								42.1	57.9		47.4	10.5			
Farm machinery																
Tractor	100.0								94.7	5.3			5.3			
Other	100.0								100.0							
Tubewell water	100.0								89.5	10.5		5.3			2.6	2.6
	25.00-50.00 Acres								Over 50.00 Acres							
Fertilizer																
Nitrogenous		100.0			2.6	31.6	63.2	2.6		100.0				26.5	38.2	35.3
Phosphatic	47.4	52.6		2.6	26.3	21.1				100.0		5.9	67.6	26.5		
Other	100.0								100.0							
New seeds																
Wheat (Mexi-Pak)	2.6	97.4			2.6	92.1		2.6		100.0		2.9	29.4	67.7		
Rice (IRRI)	92.1	7.9			7.9				79.4	20.6			17.7	2.9		
Pesticides	26.3	73.7		44.7	29.0				35.3	64.7		29.4	35.3			
Farm machinery																
Tractor	68.4	31.6		7.9	23.7					100.0	55.9	41.2		2.9		
Other	100.0								97.1	2.9			2.9			
Tubewell water	86.8	13.2		2.6	10.5				2.9	97.1		14.7	35.8	25.0	20.6	

(Rahimyar Khan)

	Under 12.50 Acres								12.50-25.00 Acres							
Fertilizer																
Nitrogenous		100.0	5.0	30.0	25.0		5.0	35.0		100.0		20.0	30.0		5.0	45.0
Phosphatic	85.0	15.0	5.0				5.0	5.0	75.0	25.0	5.0	5.0	10.0		5.0	
Other	100.0								100.0							
New seeds																
Wheat (Mexi-Pak)	25.0	75.0	10.0	50.0	5.0		10.0		15.0	85.0	10.0	35.0	30.0		10.0	
Rice (IRRI)	100.0								100.0							
Pesticides	80.0	20.0	10.0				5.0	5.0	60.0	40.0		10.0		5.0	20.0	5.0
Farm machinery																
Tractor	60.0	40.0	5.0	5.0	20.0		5.0	5.0	75.0	25.0		5.0	5.0	5.0	10.0	
Other	100.0								100.0							
Tubewell water	40.0	60.0	10.0	15.0	20.0	5.0	5.0	5.0	40.0	60.0	15.0	10.0	10.0	15.0	5.0	5.0
	25.00-50.00 Acres								Over 50.00 Acres							
Fertilizer																
Nitrogenous		100.0			30.0	10.0		60.0		100.0			28.6		7.1	64.3
Phosphatic	60.0	40.0	10.0	20.0	10.0				64.3	35.7	7.1	14.3				
Other	100.0															
New seeds																
Wheat (Mexi-Pak)	25.0	75.0	5.0	45.0	10.0	5.0	10.0		14.3	85.7		14.3	21.4	14.3	35.7	
Rice (IRRI)	100.0								100.0							
Pesticides	45.0	55.0	5.0	15.0	25.0			10.0	35.7	64.3	7.1	21.4	21.4			14.3
Farm machinery																
Tractor	60.0	40.0	5.0	20.0		5.0		10.0	28.6	71.4	7.1	21.4	7.1	21.4	7.1	7.1
Other	100.0								100.0							
Tubewell water	30.0	70.0	10.0	25.0	20.0			15.0	42.9	57.1		21.4	7.1	14.3	7.1	7.1

(continued)

TABLE A.5 (continued)
(Jacobabad)

Input	% Never Used	% Used Once	Percentage Farmers Using 1972-73	2 Yrs.	4 Yrs.	6 Yrs.	8 Yrs.	Over 8 Yrs.	% Never Used	% Used Once	Percentage Farmers Using 1972-73	2 Yrs.	4 Yrs.	6 Yrs.	8 Yrs.	Over 8 Yrs.
	Under 12.50 Acres								12.50-25.00 Acres							
Fertilizer																
Nitrogenous	18.2	81.8		18.2	63.6				8.3	91.7		41.7	50.0			
Phosphatic	100.0								100.0							
Other	100.0								100.0							
New seeds																
Wheat (Mexi-Pak)	9.1	90.9	18.2	45.5	27.3					100.0		75.0	25.0			
Rice (IRRI)		100.0		63.6	36.4					100.0		58.3	41.7			
Pesticides	100.0								100.0							
Farm machinery																
Tractor	90.9	9.1			9.1				66.7	33.3	8.3		8.3	16.7		
Other	100.0								100.0							
Tubewell water	100.0								100.0							
	25.00-50.00 Acres								Over 50.00 Acres							
Fertilizer																
Nitrogenous		100.0	25.0		75.0					100.0		45.5	36.4	18.2		
Phosphatic	91.7	8.3		8.3					81.8	18.2		9.1	9.1			
Other	100.0								100.0							
New seeds																
Wheat (Mexi-Pak)	8.3	91.7	41.7	41.7	8.3					100.0		54.6	45.5			
Rice (IRRI)	8.3	91.7	33.3	25.0	33.3					100.0		45.5	54.6			
Pesticides	100.0								90.9	9.1				9.1		
Farm machinery																
Tractor	58.3	41.7	8.3	16.7	8.3	8.3			9.1	90.9		27.3	18.2	27.3	18.2	
Other	100.0								90.9	9.1		9.1				
Tubewell water	100.0								81.8	18.2			9.1			9.1

(Larkana)

	Under 12.50 Acres							12.50-25.00 Acres						
Fertilizer														
Nitrogenous		100.0		36.4	45.5	9.1	9.1		100.0		8.3	58.3	33.3	
Phosphatic	90.9	9.1				9.1		83.3	16.7		16.7			
Other	100.0							100.0						
New seeds														
Wheat (Mexi-Pak)	81.8	18.2			18.2			41.7	58.3		41.7	16.7		
Rice (IRRI)	18.2	81.8		18.8	45.5	18.2			100.0		8.3	50.0	41.7	
Pesticides	45.5	54.5		9.1	27.3	9.1	9.1	75.0	25.0		16.7	8.3		
Farm machinery														
Tractor	90.9	9.1			9.1			41.7	58.3		33.3	16.7	8.3	
Other	100.0							100.0						
Tubewell water	100.0							100.0						
	25.00-50.00 Acres							Over 50 Acres						
Fertilizer														
Nitrogenous		100.0		16.6	75.0	8.3			100.0		9.1	54.6	36.4	
Phosphatic	66.7	33.3	8.3	16.7	8.3			72.7	27.3			27.3		
Other	100.0							100.0						
New seeds														
Wheat (Mexi-Pak)	58.3	41.7		25.0	16.7			45.5	54.5		45.5	9.1		
Rice (IRRI)		100.0		8.3	75.0	16.7			100.0		18.2	45.5	36.4	
Pesticides	58.3	41.7		16.7	8.3	8.3	8.3	45.5	54.5		27.3		27.3	
Farm machinery														
Tractor	50.0	50.0		33.3	16.7			18.2	81.8		45.5	36.4		
Other	91.7	8.3		8.3				90.9	9.1			9.1		
Tubewell water	100.0							100.0						

(continued)

TABLE A.5 (continued)
(Nawabshah)

Input	% Never Used	% Used Once	Percentage Farmers Using 1972-73	2 Yrs.	4 Yrs.	6 Yrs.	8 Yrs.	Over 8 Yrs.	% Never Used	% Used Once	Percentage Farmers Using 1972-73	2 Yrs.	4 Yrs.	6 Yrs.	8 Yrs.	Over 8 Yrs.
	Under 12.50 Acres								12.50-25.00 Acres							
Fertilizer																
Nitrogenous		100.0		42.9	50.0	7.1				100.0		21.4	57.1	21.4		
Phosphatic									92.9	7.1	7.1					
Other	92.9	7.1		7.1					100.0							
New seeds																
Wheat (Mexi-Pak)	50.0	50.0		28.6	21.4				50.0	50.0		14.3	35.7			
Rice (IRRI)	92.9	7.1			7.1				100.0							
Pesticides	21.4	78.6		21.4	57.1				7.2	92.9		28.6	57.1	7.1		
Farm machinery																
Tractor	64.3	35.7		7.1	21.4	7.1			50.0	50.0		42.9				7.1
Other	100.0								100.0							
Tubewell water	100.0								100.0							
	25.00-50.00 Acres								Over 50.00 Acres							
Fertilizer																
Nitrogenous		100.0		50.0	50.0					100.0		28.6	42.9	28.6		
Phosphatic	78.6	21.4		21.4					64.3	35.7		21.4	14.3			
Other	100.0								100.0							
New seeds																
Wheat (Mexi-Pak)	42.9	57.1		21.4	28.6	7.1			28.6	71.4		21.4	50.0			
Rice (IRRI)	100.0								100.0							
Pesticides		100.0		35.7	50.0	14.3				100.0		21.4	64.3	14.3		
Farm machinery																
Tractor	42.9	57.1	7.1	21.4	14.3	14.3			35.7	64.3		35.7	28.6			
Other	100.0								71.4	28.6		21.4	7.1			
Tubewell water	100.0								100.0							

(Hyderabad)

	Under 12.50 Acres								12.50-25.00 Acres							
Fertilizer																
Nitrogenous	9.5	90.5		19.1	35.7	28.6		19.1		100.0			33.3	33.3	9.5	23.8
Phosphatic	95.2	4.8		4.8					100.0							
Other	100.0								100.0							
New seeds																
Wheat (Mexi-Pak)	33.3	66.7	4.8	19.1	28.6	14.3			42.9	57.1			28.6	28.6		
Rice (IRRI)	52.4	47.6	4.8	28.6	14.3				52.4	46.6		4.8	38.1	4.8		
Pesticides	90.5	9.5		4.8				4.8	90.5	9.5			4.8	4.8		
Farm machinery																
Tractor	100.0								100.0							
Other	100.0								100.0							
Tubewell water	100.0								95.2	4.8		4.8				

	25.00-50.00 Acres								Over 50.00 Acres							
Fertilizer																
Nitrogenous	5.0	95.0			10.0	20.0	40.0	25.0		100.0		5.0	10.0	20.0	30.0	35.0
Phosphatic	100.0								84.0	15.0			10.0	5.0		
Other	95.0	5.0						5.0	100.0							
New seeds																
Wheat (Mexi-Pak)	40.0	60.0		5.0	45.0	10.0			25.0	75.0		5.0	30.0	25.0	5.0	10.0
Rice (IRRI)	55.0	45.0			35.0	5.0	5.0		50.0	50.0		5.0	30.0	15.0		
Pesticides	100.0								65.0	35.0		5.0	15.0	5.0	5.0	5.0
Farm machinery																
Tractor	100.0								75.0	25.0		15.0	10.0			
Other	100.0								95.0	5.0		5.0				
Tubewell water	95.0	5.0		5.0					90.0	10.0		5.0	5.0			

Note: Blank spaces indicate data not applicable.

Source: Compiled by the author.

TABLE A.6

Use of Selected Inputs
(Jhelum)

Crop	Fertilizer		Irrigation Water: Canal		Irrigation Water: Sur. Well		Irrigation Water: Tubewell		New Seeds		Farm Machinery
	Av. No. Bags/Acre	% Farmers Using	Av. No. Irr./Ac.	% Farmers Using	Av. No. Irr./Ac.	% Farmers Using	Av. No. Irr./Ac.	% Farmers Using	% Farmers Growing Any Crop	% Farmer Growing This Crop	% Farmers Using
	Under 12.50 Acres										
Wheat											
Local	0.3	14.3			2	14.3					28.6
Mexi-Pak											
Rice											
Local											
Basmati											
IRRI											
Cotton											
Local											
Improved											
Sugarcane											
	12.50-25.00 Acres										
Wheat											
Local											28.6
Mexi-Pak											
Rice											
Local											
Basmati											
IRRI											
Cotton											
Local											
Improved											
Sugarcane											

(Jhelum, continued)

	25.00-50.00 Acres									
Wheat										
Local										
Mexi-Pak										
Rice										
Local										
Basmati										
IRRI										
Cotton										
Local										
Improved										
Sugarcane										
	Over 50.00 Acres									
Wheat										
Local										100.0
Mexi-Pak	0.1	50.0						100.0	66.0	100.0
Rice										
Local										
Basmati										
IRRI										
Cotton										
Local										
Improved										
Sugarcane										

(continued)

TABLE A.6 (continued)
(Gujranwala)

Crop	Fertilizer		Irrigation Water: Canal		Irrigation Water: Sur. Well		Irrigation Water: Tubewell		New Seeds		Farm Machinery
	Av. No. Bags/Acre	% Farmers Using	Av. No. Irr./Ac.	% Farmers Using	Av. No. Irr./Ac.	% Farmers Using	Av. No. Irr./Ac.	% Farmers Using	% Farmers Growing Any Crop	% Farmers Growing This Crop	% Farmers Using
	Under 12.50 Acres										
Wheat											
Local			4	100.0			1	100.0			
Mexi-Pak	0.7	53.6	5	60.7			4	50.0	100.0	96.6	25.0
Rice											
Local	1.0	100.0	14	50.0			10	100.0			
Basmati	0.5	92.3	15	61.5			13	46.2			15.4
IRRI	0.5	96.4	16	57.1			14	50.0	100.0	50.0	25.0
Cotton											
Local											
Improved											
Sugarcane	2.0	37.5	15	87.5			6	37.5			12.5
	12.50-25.00 Acres										
Wheat											
Local	1.0	100.0	5	100.0	2	50.0					
Mexi-Pak	1.0	100.0	5	100.0	2	12.5			100.0	92.3	4.2
Rice											
Local	1.0	100.0	12	100.0	4	100.0					
Basmati	1.9	95.5	11	100.0	4	54.5					4.5
IRRI	1.2	100.0	10	100.0	3	47.6			87.5	46.7	4.8
Cotton											
Local	2.0	100.0	6	100.0							
Improved											
Sugarcane	1.7	83.3	20	100.0	4	66.7					

(Gujranwala, continued)

	25.00-50.00 Acres									
Wheat										
Local	1.0	100.0	5	100.0			2	50.0			
Mexi-Pak	1.4	100.0	5	100.0	3	3.4	2	51.7	100.0	93.3	10.3
Rice											
Local	1.0	100.0	12	100.0	4	10.0	4	90.0			
Basmati	1.4	100.0	11	100.0	6	3.8	5	92.3			11.5
IRRI	1.0	100.0	11	100.0			4	88.0	89.3	41.1	12.0
Cotton											
Local	1.9	50.0	5	100.0			4	50.0			
Improved											
Sugarcane	3.1	75.0	17	100.0			5	83.3			16.7
	Over 50.00 Acres										
Wheat											
Local	1.0	75.0	4	100.0			3	100.0			50.0
Mexi-Pak	1.0	100.0	4	91.7			3	100.0	100.0	7.5	75.0
Rice											
Local	1.0	100.0	11	80.0			9	100.0			80.0
Basmati	1.0	100.0	12	83.3			10	100.0			75.0
IRRI	1.0	100.0	10	91.7			10	100.0	100.0	41.4	75.0
Cotton											
Local											
Improved											
Sugarcane	1.1	83.3	16	100.0			7	100.0			50.0

(continued)

TABLE A.6 (continued)
(Sahiwal)

Crop	Fertilizer		Irrigation Water: Canal		Irrigation Water: Sur. Well		Irrigation Water: Tubewell		New Seeds		Farm Machinery
	Av. No. Bags/Acre	% Farmers Using	Av. No. Irr./Ac.	% Farmers Using	Av. No. Irr./Ac.	% Farmers Using	Av. No. Irr./Ac.	% Farmers Using	% Farmers Growing Any Crop	% Farmers Growing This Crop	% Farmers Using
	Under 12.50 Acres										
Wheat											
Local	1.6	100.0	4	100.0			2	33.3			8.3
Mexi-Pak	1.7	97.7	5	100.0			2	16.3	93.5	78.2	18.6
Rice											
Local	2.0	100.0	13	100.0			6	50.0			
Basmati	1.9	100.0	16	100.0			8	26.5			8.8
IRRI	1.0	100.0	6	100.0					2.2	2.5	
Cotton											
Local	0.3	25.0	10	100.0			4	25.0			7.5
Improved	1.5	50.0	9	100.0	1	2.4	4	14.3	91.3	93.3	14.3
Sugarcane	1.5	92.5	21	100.0	1	2.5	7	25.0			20.0
	12.50-25.00 Acres										
Wheat											
Local	1.6	100.0	4	100.0	1	4.8	2	61.9			4.8
Mexi-Pak	1.6	100.0	4	100.0	1	2.2	3	58.7	100.0	67.7	10.9
Rice											
Local	1.7	100.0	13	100.0	10	12.5	7	87.5			
Basmati	2.2	100.0	14	100.0	8	5.1	9	66.7			10.3
IRRI	2.2	100.0	10	100.0	6	25.0	9	50.0	8.7	7.8	
Cotton											
Local	0.2	33.3	10	100.0			7	50.0			16.7
Improved	1.5	50.0	8	100.0	5	2.5	5	65.0	87.0	87.1	10.0
Sugarcane	2.3	93.5	21	100.0	8		9	60.9			10.9

(Sahiwal, continued)

	25.00-50.00 Acres								
Wheat									
Local	1.5	100.0	4	100.0	3	92.6			22.2
Mexi-Pak	1.6	100.0	4	100.0	3	93.5	100.0	60.0	28.3
Rice									
Local	0.8	100.0	12	100.0	9	100.0			66.7
Basmati	1.8	100.0	13	100.0	9	95.1			26.8
IRRI	1.6	100.0	12	100.0	9	100.0	32.6	25.4	13.3
Cotton									
Local			8	100.0	4	100.0			
Improved	1.4	57.1	7	100.0	5	90.5	91.3	91.3	28.6
Sugarcane	1.8	97.7	19	100.0	9	93.0			25.6
	Over 50.00 Acres								
Wheat									
Local	1.3	100.0	3	100.0	3	100.0			100.0
Mexi-Pak	1.7	100.0	4	100.0	3	95.7	100.0	75.4	95.7
Rice									
Local	1.0	100.0	10	100.0	9	100.0			100.0
Basmati	2.3	97.6	12	100.0	10	97.6			95.1
IRRI	2.2	100.0	9	100.0	10	100.0	10.9	10.0	100.0
Cotton									
Local	1.0	100.0	9	100.0	6	100.0			100.0
Improved	1.2	82.2	7	100.0	6	95.6	97.8	97.8	95.6
Sugarcane	2.4	97.4	19	100.0	10	94.7			94.7

(continued)

TABLE A.6 (continued)
(Lyallpur)

	Fertilizer		Irrigation Water						New Seeds		Farm Machinery
			Canal		Sur. Well		Tubewell				
Crop	Av. No. Bags/Acre	% Farmers Using	Av. No. Irr./Ac.	% Farmers Using	Av. No. Irr./Ac.	% Farmers Using	Av. No. Irr./Ac.	% Farmers Using	% Farmers Growing Any Crop	% Farmers Growing This Crop	% Farmers Using
					Under 12.50 Acres						
Wheat											
Local											
Mexi-Pak	0.9	100.0	5	100.0			1	2.6	100.0	100.0	2.6
Rice											
Local			17	100.0							
Basmati	1.3	50.0	15	100.0							
IRRI											
Cotton											
Local											
Improved	0.7	94.7	6	100.0			5	5.3	100.0	100.0	2.6
Sugarcane	3.2	100.0	16	100.0			5	5.3			2.6
					12.50-25.00 Acres						
Wheat											
Local											
Mexi-Pak	1.5	100.0	5	100.0			2	10.5	100.0	100.0	5.3
Rice											
Local	1.9	100.0	17	100.0							
Basmati	1.0	100.0	17	100.0			1	12.5			
IRRI											
Cotton											
Local											
Improved	1.0	100.0	6	100.0			6	2.6	100.0	100.0	2.6
Sugarcane	3.3	100.0	16	100.0			5	7.9			5.3

(Lyallpur, continued)

	25.00-50.00 Acres								
Wheat									
Local									
Mexi-Pak	2.1	100.0	5	100.0	2	15.8	100.0	100.0	28.9
Rice									
Local	1.3	50.0	18	100.0					
Basmati	0.6	100.0	16	100.0	2	16.7			50.0
IRRI									
Cotton									
Local									
Improved	0.9	100.0	6	100.0	5	13.2	100.0	100.0	28.9
Sugarcane	2.7	100.0	16	100.0	6	15.8			31.6
	Over 50.00 Acres								
Wheat									
Local									
Mexi-Pak	2.0	100.0	4	100.0	2	67.6	100.0	100.0	97.1
Rice									
Local									
Basmati	1.4	100.0	15	100.0	4	100.0			100.0
IRRI									
Cotton									
Local									
Improved	1.1	100.0	6	100.0	3	87.5	94.1	100.0	96.9
Sugarcane	2.0	97.1	16	100.0	4	97.1			97.1

(continued)

TABLE A.6 (continued)
(Rahimyar Khan)

Crop	Fertilizer		Irrigation Water: Canal		Irrigation Water: Sur. Well		Irrigation Water: Tubewell		New Seeds		Farm Machinery
	Av. No. Bags/Acre	% Farmers Using	Av. No. Irr./Ac.	% Farmers Using	Av. No. Irr./Ac.	% Farmers Using	Av. No. Irr./Ac.	% Farmers Using	% Farmers Growing Any Crop	% Farmers Growing This Crop	% Farmers Using
	Under 12.50 Acres										
Wheat											
Local	0.8	88.9	5	22.2			4	88.9			
Mexi-Pak	2.5	86.7	5	53.3	4	13.3	4	46.7	75.0	62.5	20.0
Rice											
Local											
Basmati											
IRRI											
Cotton											
Local	0.9	42.9	5	100.0			2	42.9			
Improved	2.4	91.7	5	100.0	4	8.3	3	33.3	60.0	63.2	8.3
Sugarcane	3.1	71.4	14	100.0			4	28.6			7.1
	12.50–25.00 Acres										
Wheat											
Local	1.2	83.3	5	33.3	3	16.7	4	66.7			
Mexi-Pak	1.9	93.3	4	46.7			5	60.0	75.0	71.4	20.0
Rice											
Local			15	100.0							
Basmati											
IRRI	2.0	100.0	4	100.0			2	100.0	5.0	50.0	
Cotton											
Local	0.9	66.7	2	100.0			3	66.7			
Improved	2.6	82.4	4	100.0			3	58.8	85.0	85.0	11.8
Sugarcane	2.0	100.0	7	100.0			4	73.3			13.3

(Rahimyar Khan, continued)

	25.00-50.00 Acres									
Wheat										
Local	1.0	100.0	4	25.0			5	75.0			37.5
Mexi-Pak	2.4	86.7	4	40.0			5	66.7	70.0	63.6	53.3
Rice										
Local	2.4	33.3	14	66.7			6	33.3			33.3
Basmati	1.0	100.0	7	100.0							
IRRI											
Cotton											
Local	1.5	100.0	3	100.0			4	100.0			100.0
Improved	2.1	94.7	4	100.0			3	47.4	90.0	94.7	42.1
Sugarcane	3.9	89.5	9	100.0			6	42.1			31.6
	Over 50.00 Acres										
Wheat											
Local	1.0	100.0			3	100.0					100.0
Mexi-Pak	3.6	92.3	4	61.5			5	53.8	92.9	92.9	53.8
Rice											
Local			6	100.0			6	33.3			33.3
Basmati											
IRRI											
Cotton											
Local	1.7	100.0	3	100.0			2	66.7			66.7
Improved	2.3	90.9	4	100.0			3	45.5	78.6	78.6	54.5
Sugarcane	3.4	92.9	10	100.0			6	42.9			35.7

(continued)

TABLE A.6 (continued)
(Jacobabad)

Crop	Fertilizer		Irrigation Water: Canal		Irrigation Water: Sur. Well.		Irrigation Water: Tubewell		New Seeds		Farm Machinery
	Av. No. Bags/Acre	% Farmers Using	Av. No. Irr./Ac.	% Farmers Using	Av. No. Irr./Ac.	% Farmers Using	Av. No. Irr./Ac.	% Farmers Using	% Farmers Growing Any Crop	% Farmers Growing This Crop	% Farmers Using
	Under 12.50 Acres										
Wheat											
Local			5	100.0							
Mexi-Pak											
Rice											
Local	0.2	36.4	20	100.0			12	9.1			
Basmati											
IRRI	0.5	100.0	20	100.0							
Cotton											
Local											
Improved											
Sugarcane											
	12.50-25.00 Acres										
Wheat											
Local			5	100.0							
Mexi-Pak			5	100.0					33.3	57.1	
Rice											
Local	0.3	45.5	20	100.0							9.1
Basmati											
IRRI	0.8	100.0	22	100.0					8.3	8.3	100.0
Cotton											
Local											
Improved											
Sugarcane											

(Jacobabad, continued)

	25.00-50.00 Acres								
Wheat									
Local			5	100.0					
Mexi-Pak			11	100.0			66.7	80.0	
Rice									
Local	0.4	63.6	19	100.0					
Basmati									
IRRI	0.8	66.7	23	100.0			16.7	18.2	
Cotton									
Local									
Improved									
Sugarcane									
	Over 50.00 Acres								
Wheat									
Local			5	80.0	2	20.0			20.0
Mexi-Pak			5	77.8	3	22.2	75.0	64.3	11.1
Rice									
Local	0.4	90.9	20	81.8	21	18.2			18.2
Basmati									
IRRI	0.4	100.0	22	75.0	25	25.0	33.3	26.3	75.0
Cotton									
Local									
Improved							8.3	100.0	
Sugarcane									

(continued)

TABLE A.6 (continued)
(Larkana)

Crop	Fertilizer Av. No. Bags/Acre	Fertilizer % Farmers Using	Irrigation Water: Canal Av. No. Irr./Ac.	Canal % Farmers Using	Sur. Well Av. No. Irr./Ac.	Sur. Well % Farmers Using	Tubewell Av. No. Irr./Ac.	Tubewell % Farmers Using	New Seeds % Farmers Growing Any Crop	New Seeds % Farmers Growing This Crop	Farm Machinery % Farmers Using
	Under 12.50 Acres										
Wheat											
Local			3	100.0							
Mexi-Pak											
Rice											
Local											
Basmati			30	100.0							
IRRI	1.4	100.0	33	100.0					100.0	100.0	
Cotton											
Local											
Improved											
Sugarcane											
	12.50-25.00 Acres										
Wheat											
Local			3	100.0							
Mexi-Pak											
Rice											
Local											
Basmati											
IRRI	1.5	100.0	35	100.0					100.0	100.0	
Cotton											
Local											
Improved											
Sugarcane											

(Larkana, continued)

	25.00-50.00 Acres									
Wheat										
Local	0.3	8.3	2	100.0	3	8.3				
Mexi-Pak										
Rice										
Local	1.0	100.0	15	100.0						
Basmati	1.0	100.0	22	100.0						
IRRI	1.4	100.0	36	100.0				100.0	100.0	
Cotton										
Local										
Improved										
Sugarcane										
	Over 50.00 Acres									
Wheat										
Local	0.3	41.7	2	50.0						
Mexi-Pak										
Rice										
Local	0.4	100.0	30	100.0						
Basmati	0.4	100.0	31	100.0						
IRRI	1.4	100.0	36	100.0				100.0	70.6	8.3
Cotton										
Local										
Improved										
Sugarcane										

(continued)

TABLE A.6 (continued)
(Nawabshah)

Crop	Fertilizer		Irrigation Water: Canal		Irrigation Water: Sur. Well		Irrigation Water: Tubewell		New Seeds		Farm Machinery
	Av. No. Bags/Acre	% Farmers Using	Av. No. Irr./Ac.	% Farmers Using	Av. No. Irr./Ac.	% Farmers Using	Av. No. Irr./Ac.	% Farmers Using	% Farmers Growing Any Crop	% Farmers Growing This Crop	% Farmers Using
	Under 12.50 Acres										
Wheat											
Local	0.4	12.5	1	100.0							
Mexi-Pak	1.7	100.0	9	100.0					28.6	33.3	
Rice											
Local											
Basmati											
IPRI											
Cotton											
Local	1.4	100.0	11	100.0							
Improved	1.5	100.0	10	100.0					57.1	72.7	
Sugarcane	1.6	100.0	20	100.0							
	12.50-25.00 Acres										
Wheat											
Local	0.2	25.0	1	100.0							
Mexi-Pak	1.6	100.0	10	100.0					28.6	50.0	
Rice											
Local											
Basmati											
IRRI											
Cotton											
Local	1.3	100.0	11	100.0							
Improved	1.7	100.0	11	100.0					57.1	72.7	
Sugarcane	1.8	100.0	23	100.0							

(Nawabshah, continued)

	25.00-50.00 Acres								
Wheat									
Local			2	100.0					
Mexi-Pak	1.6	100.0	9	100.0			35.7	45.5	
Rice									
Local									
Basmati									
IRRI									
Cotton									
Local									
Improved	1.4	100.0	10	100.0			78.6	100.0	
Sugarcane	1.8	100.0	22	100.0					
	Over 50.00 Acres								
Wheat									
Local	0.7	12.5	3	100.0	2	12.5			
Mexi-Pak	1.6	100.0	10	80.0	3	20.0	35.7	38.5	20.0
Rice									
Local									
Basmati									
IRRI									
Cotton									
Local	1.5	100.0	12	100.0					
Improved	1.7	100.0	10	91.7	7	16.7	85.7	92.3	8.3
Sugarcane	1.8	100.0	23	91.7	13	16.7			8.3

(continued)

TABLE A.6 (continued)
(Hyderabad)

Crop	Fertilizer		Irrigation Water: Canal		Irrigation Water: Sur. Well		Irrigation Water: Tubewell		New Seeds		Farm Machinery
	Av. No. Bags/Acre	% Farmers Using	Av. No. Irr./Ac.	% Farmers Using	Av. No. Irr./Ac.	% Farmers Using	Av. No. Irr./Ac.	% Farmers Using	% Farmers Growing Any Crop	% Farmers Growing This Crop	% Farmers Using
					Under 12.50 Acres						
Wheat											
Local											
Mexi-Pak	1.3	68.4	6	100.0			1	5.3	86.4	100.0	5.3
Rice											
Local	0.9	100.0	22	100.0							
Basmati											
IRRI	0.7	80.0	15	100.0					45.5	83.3	
Cotton											
Local	0.5	25.0	8	100.0							
Improved	1.4	80.0	6	100.0					45.5	71.4	10.0
Sugarcane	3.7	100.0	9	100.0							
					12.50-25.00 Acres						
Wheat											
Local											
Mexi-Pak	2.3	76.5	6	100.0			3	5.9	77.3	100.0	
Rice											
Local	0.9	100.0	25	100.0							
Basmati											
IRRI	0.8	100.0	16	100.0					45.5	83.3	
Cotton											
Local			6	100.0							
Improved	2.0	81.8	6	100.0			3	9.1	50.0	91.7	
Sugarcane	1.0	87.5	11	100.0							

(Hyderabad, continued)

	25.00-50.00 Acres								
Wheat									
Local									
Mexi-Pak	1.7	75.0	7	100.0	2	6.3	80.0	100.0	
Rice									
Local	1.0	100.0	21	100.0					
Basmati									
IRRI	1.4	100.0	15	100.0			45.0	81.8	
Cotton									
Local	1.0	50.0	9	100.0					
Improved	1.7	77.8	7	100.0	1	11.1	45.0	81.8	
Sugarcane	1.4	100.0	11	100.0	1	9.1			
	Over 50.00 Acres								
Wheat									
Local									
Mexi-Pak	2.6	87.5	6	100.0	5	18.8	80.0	100.0	31.3
Rice									
Local	1.1	100.0	17	100.0					33.3
Basmati									
IRRI	1.4	100.0	15	100.0			50.0	76.9	20.0
Cotton									
Local	2.1	100.0	9	100.0					33.3
Improved	1.4	100.0	7	100.0	4	33.3	45.0	75.0	44.4
Sugarcane	1.8	100.0	10	100.0	6	7.1			28.6

Note: Blank spaces indicate data not applicable.

Source: Compiled by the author.

TABLE A.7

Distribution of Average Value of Farm Output by Crop
(rupees)

Crop	Under 12.50 Acres		12.50-25.00 Acres		25.00-50.00 Acres		Over 50.00 Acres	
	Amount	(%)	Amount	(%)	Amount	(%)	Amount	(%)
				Jhelum				
Wheat								
Local	424.8	34.0	622.9	18.7	1,786.0	40.5	3,800.0	30.1
Mexi-Pak							1,995.0	31.6
Rice								
Local								
Basmati								
IRRI								
Cotton								
Local								
Improved								
Sugarcane								
Maize								
Local			70.0	0.3	119.0	0.9		
Improved								
				Gujranwala				
Wheat								
Local	190.0	0.2	1,200.0	0.7	1,995.0	0.5	2,272.5	1.5
Mexi-Pak	911.9	23.0	3,921.5	27.6	9,836.4	35.4	19,584.2	38.4
Rice								
Local	537.5	1.0	400.0	0.2	3,544.5	4.4	5,094.0	4.2
Basmati	1,390.4	32.5	4,574.2	29.5	7,461.7	24.0	9,382.5	18.4
IRRI	866.5	21.8	1,867.6	11.5	6,040.4	18.7	14,007.0	27.4
Cotton								
Local			210.0	0.0	1,875.0	0.5		
Improved								
Sugarcane	945.0	6.8	1,715.0	3.0	1,860.4	2.8	1,219.1	1.2
Maize								
Local			160.0	0.0				
Improved								
				Sahiwal				
Wheat								
Local	990.8	2.7	1,163.2	2.5	1,555.4	2.2	1,854.0	0.7
Mexi-Pak	1,519.4	14.9	3,431.0	16.0	7,027.6	17.2	16,843.6	19.0
Rice								
Local	842.5	0.8	1,223.1	1.0	1,483.3	0.2	3,702.5	0.4
Basmati	1,471.5	11.4	2,705.0	10.7	4,528.9	9.9	16,181.2	16.3
IRRI	600.0	0.1	1,316.8	0.5	2,168.2	1.7	6,180.0	0.8
Cotton								
Local	1,902.5	1.7	3,697.5	2.3	5,905.0	1.3	18,900.0	0.5
Improved	3,312.9	31.8	7,643.3	31.0	14,004.1	31.4	31,645.6	35.0
Sugarcane	2,091.1	19.1	3,149.4	14.7	5,453.0	12.5	7,915.0	7.4
Maize								
Local	75.0	0.0						
Improved	354.4	1.8	690.3	2.5	937.0	1.9	1,973.3	1.6

TABLE A.7 (continued)

Crop	Under 12.50 Acres		12.50-25.00 Acres		25.00-50.00 Acres		Over 50.00 Acres	
	Amount	(%)	Amount	(%)	Amount	(%)	Amount	(%)
				Lyallpur				
Wheat								
Local								
Mexi-Pak	1,619.6	21.8	3,583.4	21.4	8,669.4	19.7	24,719.0	19.4
Rice								
Local	108.0	0.0	189.0	0.0	490.0	0.0		
Basmati	342.0	0.2	577.5	0.7	1,132.9	0.4	2,752.9	0.6
IRRI								
Cotton								
Local								
Improved	2,233.2	30.0	5,553.2	33.1	14,570.8	33.2	47,371.1	35.0
Sugarcane	1,381.2	18.6	3,859.6	23.0	13,922.7	31.7	40,022.6	31.5
Maize								
Local	99.6	0.2	193.3	0.0				
Improved	371.1	3.7	830.8	4.3	1,383.1	3.0	2,799.3	1.7
				Rahimyar Khan				
Wheat								
Local	497.6	3.9	1,100.0	2.6	1,556.7	2.8	2,812.5	0.3
Mexi-Pak	967.2	12.6	2,074.7	12.2	2,628.2	8.7	10,331.8	11.8
Rice								
Local			900.0	0.4	630.0	0.4	740.0	0.2
Basmati					2,250.0	0.5		
IRRI			486.0	0.2				
Cotton								
Local	1,667.1	10.1	4,213.3	5.0	18,000.0	4.0	21,383.3	5.6
Improved	3,244.4	33.8	5,817.9	38.9	9,134.2	38.3	42,480.5	41.0
Sugarcane	1,564.4	19.0	4,242.4	25.0	7,706.1	32.3	23,100.3	28.4
Maize								
Local	96.0	0.0	192.0	0.0	96.0	0.0	64.0	0.0
Improved								
				Jacobabad				
Wheat								
Local	1,562.5	10.6	1,834.7	4.2	2,364.0	1.9	5,348.0	3.9
Mexi-Pak			1,305.0	4.0	1,562.6	5.1	5,512.1	7.3
Rice								
Local	3,506.1	65.5	6,645.6	55.6	11,781.1	53.0	26,585.6	42.8
Basmati								
IRRI	1,320.0	2.2	14,168.0	10.8	11,730.7	14.4	26,323.0	15.4
Cotton								
Local								
Improved								
Sugarcane								
Maize								
Local								
Improved								

(continued)

TABLE A.7 (continued)

Crop	Under 12.50 Acres		12.50-25.00 Acres		25.00-50.00 Acres		Over 50.00 Acres	
	Amount	(%)	Amount	(%)	Amount	(%)	Amount	(%)
				Larkana				
Wheat								
Local	1,309.2	8.7	1,758.3	5.9	3,491.7	8.4	6,823.3	8.1
Mexi-Pak								
Rice								
Local					1,920.0	0.4	1,508.0	0.5
Basmati	1,020.0	1.1			2,282.5	0.9	3,750.0	0.7
IRRI	5,853.5	71.5	13,557.9	60.4	24,402.8	58.8	41,974.2	50.0
Cotton								
Local								
Improved								
Sugarcane								
Maize								
Local								
Improved								
				Nawabshah				
Wheat								
Local	2,312.9	9.6	2,572.0	1.9	5,130.7	2.9	8,274.4	2.8
Mexi-Pak	5,772.5	11.9	10,975.0	8.1	15,810.0	7.5	29,290.0	6.1
Rice								
Local								
Basmati								
IRRI								
Cotton								
Local	6,570.0	10.2	4,418.7	2.5			45,000.0	1.9
Improved	8,773.0	36.3	10,469.8	15.5	22,045.4	23.1	42,988.2	21.6
Sugarcane	11,293.3	17.5	40,700.0	60.2	72,575.6	62.3	113,626.3	57.1
Maize								
Local							12,000.0	0.5
Improved								
				Hyderabad				
Wheat								
Local								
Mexi-Pak	1,493.5	22.6	2,645.5	17.6	3,429.6	13.8	12,575.3	18.7
Rice								
Local	715.0	1.1	937.5	0.7	1,570.0	0.8	2,816.7	0.8
Basmati								
IRRI	2,862.4	22.8	4,025.9	15.8	7,607.8	17.2	12,191.8	11.4
Cotton								
Local	1,467.3	4.7	5,280.0	2.0	2,870.0	1.4	10,750.0	3.0
Improved	2,128.6	17.0	5,237.9	22.6	7,825.6	17.7	29,747.2	24.9
Sugarcane	6,305.0	20.1	9,080.6	28.4	14,709.1	40.7	17,520.7	22.8
Maize								
Local	880.0	0.7	300.0	0.1				
Improved								

Note: Blank spaces indicate data not applicable.
Source: Compiled by the author.

TABLE A.8

Distribution of Average Value of Farm Output per Acre by Crop
(rupees)

Crop	Under 12.50 Acres	12.50-25.00 Acres	25.00-50.00 Acres	Over 50.00 Acres	Under 12.50 Acres	12.50-25.00 Acres	25.00-50.00 Acres	Over 50.00 Acres	Under 12.50 Acres	12.50-25.00 Acres	25.00-50.00 Acres	Over 50.00 Acres
	Jhelum				Gujranwala				Sahiwal			
Wheat												
Local	110.1	83.1	107.1	190.0	190.0	200.0	285.0	363.6	396.3	375.2	382.2	392.0
Mexi-Pak				99.8	299.0	398.9	411.1	482.6	435.4	531.1	570.0	590.2
Rice												
Local					537.5	400.0	389.5	509.4	421.3	407.7	404.2	
Basmati					681.6	516.3	621.8	771.0	704.1	747.2	742.4	1,164.1
IRRI					470.9	377.3	418.3	492.9	600.0	585.2	590.8	657.5
Cotton												
Local						210.0	441.2		761.0	739.5	787.3	945.0
Improved									1,274.2	1,447.6	1,420.3	
Sugarcane					1,890.0	1,465.8	1,442.2	1,325.2	1,548.9	1,464.8	1,603.8	1,439.1
Maize												
Local		538.5	476.0			160.0			150.0			
Improved									302.0	298.8	282.2	310.3
	Lyallpur				Rahimyar Khan				Jacobabad			
Wheat												
Local					203.9	220.0	194.6	281.3	215.5	220.3	168.9	272.9
Mexi-Pak	456.2	481.0	562.2	596.1	343.0	334.6	297.6	372.1		174.0	171.2	232.4
Rice												
Local						900.0	377.3	105.7	385.7	354.8	361.9	391.5
Basmati	456.0	577.5	678.4	620.0			1,125.0					
IRRI		356.6	326.7			972.0			660.0	644.0	541.3	531.8
Cotton												
Local					754.4	549.3	818.2	668.2				
Improved	778.1	900.0	1,135.7	1,483.1	847.1	936.9	899.0	1,022.4				
Sugarcane	1,453.9	1,575.4	2,084.2	2,364.0	1,629.6	1,650.7	1,893.4	1,806.1				
Maize												
Local	166.0				96.0	192.0	192.0	128.0				
Improved	522.6	597.7	668.2	596.9								

(continued)

TABLE A.8 (continued)

Crop	Under 12.50 Acres	12.50-25.00 Acres	25.00-50.00 Acres	Over 50.00 Acres	Under 12.50 Acres	12.50-25.00 Acres	25.00-50.00 Acres	Over 50.00 Acres	Under 12.50 Acres	12.50-25.00 Acres	25.00-50.00 Acres	Over 50.00 Acres
	Larkana				Nawabshah				Hyderabad			
Wheat												
Local	146.1	158.7	197.6	259.9	297.3	305.5	276.0	305.0				
Mexi-Pak					881.3	707.6	752.9	898.5	394.1	428.1	332.7	454.1
Rice												
Local			480.0	564.8					317.8	288.5	261.7	211.3
Basmati	510.0		760.8	833.3								
IRRI	750.4	702.9	687.2	730.0					493.5	374.5	422.7	409.1
Cotton												
Local					985.0	698.1		1,800.0	733.6	660.0	717.5	816.3
Improved					1,271.5	1,106.7	1,095.7	1,289.8	734.0	873.0	902.6	889.6
Sugarcane					2,987.7	3,060.2	2,979.3	2,932.3	2,584.0	3,026.9	2,567.0	2,416.7
Maize												
Local								1,200.0	880.0	300.0		
Improved												

Note: Blank spaces indicate data not applicable.

Source: Compiled by the author.

TABLE A.9

Distribution of Average Variable Cost by Crop
(rupees)

Crop	Under 12.50 Acres Amount	(%)	12.50-25.00 Acres Amount	(%)	25.00-50.00 Acres Amount	(%)	Over 50.00 Acres Amount	(%)
				Jhelum				
Wheat								
Local	158.3	52.4	296.0	59.3	371.3	42.0	450.0	8.9
Mexi-Pak							1,552.5	61.7
Rice								
Local								
Basmati								
IRRI								
Cotton								
Local								
Improved								
Sugarcane								
Maize								
Local			13.2	0.4	18.0	0.7		
Improved								
				Gujranwala				
Wheat								
Local	72.4	0.4	702.4	2.0	967.8	1.0	971.3	2.1
Mexi-Pak	259.5	35.9	1,103.0	38.0	2,822.5	41.3	6,350.8	40.9
Rice								
Local	92.5	0.9	113.5	0.3	988.2	5.0	1,614.5	4.3
Basmati	176.4	22.7	903.1	28.5	1,419.3	18.6	2,063.3	13.3
IRRI	178.1	24.7	522.4	15.7	1,616.1	20.4	4,380.3	28.2
Cotton								
Local			148.8	0.2	493.5	0.5		
Improved								
Sugarcane	233.0	9.2	438.3	3.8	660.7	4.0	751.7	2.4
Maize								
Local			127.6	0.2				
Improved								
				Sahiwal				
Wheat								
Local	299.5	4.6	330.2	4.4	486.2	4.4	638.4	1.5
Mexi-Pak	423.9	23.2	823.7	24.2	1,699.0	26.4	4,392.2	32.5
Rice								
Local	228.8	1.2	327.2	1.7	464.3	0.5	1,309.4	0.8
Basmati	235.6	10.2	425.7	10.6	775.7	10.7	1,927.4	12.7
IRRI	127.0	0.2	354.8	0.9	500.6	2.5	1,524.8	1.2
Cotton								
Local	365.1	1.9	432.8	1.7	596.0	0.8	2,811.0	0.5
Improved	345.5	18.5	581.0	14.8	1,048.8	14.9	2,732.3	19.8
Sugarcane	483.1	24.6	771.3	22.6	1,184.3	17.2	1,726.2	10.5
Maize								
Local								
Improved	97.7	2.6	181.0	4.2	317.5	4.0	723.0	3.8

(continued)

TABLE A.9 (continued)

Crop	Under 12.50 Acres		12.50-25.00 Acres		25.00-50.00 Acres		Over 50.00 Acres	
	Amount	(%()	Amount	(%)	Amount	(%)	Amount	(%)
	Lyallpur							
Wheat								
Local								
Mexi-Pak	386.8	29.7	840.0	28.5	2,357.6	28.5	9,007.9	34.8
Rice								
Local	11.0	0.0	89.9	0.2	221.3	0.1		
Basmati	69.5	0.3	135.8	1.0	195.2	0.4	968.4	1.0
IRRI								
Cotton								
Local								
Improved	407.1	31.3	805.1	27.4	1,875.7	22.7	6,745.3	24.5
Sugarcane	361.1	27.8	905.1	30.7	2,930.9	35.4	6,569.4	25.4
Maize								
Local	23.7	0.2	48.0	0.1				
Improved	72.7	4.1	124.3	3.7	277.6	3.2	753.5	2.2
	Rahimyar Khan							
Wheat								
Local	303.6	10.9	534.0	6.3	943.6	7.2	1,459.0	0.4
Mexi-Pak	403.1	24.1	882.1	25.9	1,431.2	20.2	6,763.8	23.3
Rice								
Local			89.4	0.2	206.4	0.6	469.5	0.4
Basmati					228.8	0.2		
IRRI			80.2	0.2				
Cotton								
Local	178.8	5.0	569.6	3.4	3,116.6	3.0	5,618.3	4.5
Improved	518.9	24.8	767.1	25.6	1,449.6	26.3	12,761.9	37.3
Sugarcane	454.3	25.4	1,025.0	30.1	1,901.0	34.5	6,246.0	23.2
Maize								
Local			62.0	0.1	4.0		31.5	0.0
Improved								
	Jacobabad							
Wheat								
Local	225.5	12.5	277.4	4.5	434.0	3.1	752.8	3.6
Mexi-Pak			214.0	4.6	245.8	7.0	913.8	7.9
Rice								
Local	349.9	53.2	811.6	48.4	116.7	43.6	4,077.6	43.0
Basmati								
IRRI	143.6	2.0	1,898.0	10.3	1,329.8	14.2	4,694.8	18.0
Cotton								
Local								
Improved								
Sugarcane								
Maize								
Local								
Improved								

TABLE A. 9 (continued)

Crop	Under 12.50 Acres Amount	(%)	12.50-25.00 Acres Amount	(%)	25.00-50.00 Acres Amount	(%)	Over 50.00 Acres Amount	(%)
				Larkana				
Wheat								
Local	293.3	13.0	360.6	9.1	644.0	12.1	1,105.9	12.5
Mexi-Pak								
Rice								
Local					276.0	0.4	138.5	0.4
Basmati	45.0	0.3			248.0	0.8	293.5	0.5
IRRI	907.3	74.0	2,234.9	75.4	3,776.6	70.8	5,978.0	67.6
Cotton								
Local								
Improved								
Sugarcane								
Maize								
Local								
Improved								
				Nawabshah				
Wheat								
Local	330.3	12.5	361.8	1.1	522.8	1.5	1,049.1	2.1
Mexi-Pak	931.1	17.7	2,027.8	6.2	2,703.1	6.3	4,688.2	5.7
Rice								
Local								
Basmati								
IRRI								
Cotton								
Local	687.1	9.8	572.3	1.3	2,029.0		2,635.0	0.6
Improved	788.7	29.9	1,170.4	7.2		10.5	4,165.2	12.2
Sugarcane	1,220.3	17.4	12,789.0	78.3	18,578.9	78.5	24,991.9	73.4
Maize								
Local							590.0	0.1
Improved								
				Hyderabad				
Wheat								
Local								
Mexi-Pak	294.0	30.0	624.8	25.1	1,039.7	21.2	3,839.3	30.6
Rice								
Local	158.0	1.7	234.0	1.1	363.5	0.9	1,315.0	2.0
Basmati								
IRRI	347.2	18.6	740.6	17.5	1,777.8	20.4	2,745.4	13.7
Cotton								
Local	44.6	1.0	195.0	0.5	228.5	0.6	1,677.8	2.5
Improved	288.8	15.5	522.0	14.4	867.5	9.9	3,888.2	17.4
Sugarcane	1,424.1	30.6	1,913.8	36.2	3,028.3	42.4	3,609.6	25.1
Maize								
Local	33.0	0.2	26.0	0.1				
Improved								

Note: Blank spaces indicate data not applicable.

Source: Compiled by the author.

TABLE A.10

Distribution of Average Variable Cost per Acre, by Crop
(rupees)

Crop	Under 12.50 Acres	12.50-25.00 Acres	25.00-50.00 Acres	Over 50.00 Acres	Under 12.50 Acres	12.50-25.00 Acres	25.00-50.00 Acres	Over 50.00 Acres	Under 12.50 Acres	12.50-25.00 Acres	25.00-50.00 Acres	Over 50.00 Acres
	Jhelum				Gujranwala				Sahiwal			
Wheat												
Local	41.0	39.5	22.3	22.5	72.4	117.1	138.3	155.4	119.8	106.5	119.5	135.0
Mexi-Pak				77.6	85.1	112.2	118.0	156.5	121.5	127.5	137.8	153.9
Rice												
Local					92.5	113.5	108.6	161.5	114.4	109.1	126.5	130.9
Basmati					86.5	101.9	118.3	169.5	112.7	117.6	127.2	138.7
IRRI					96.8	105.5	111.9	154.1	127.0	157.7	136.4	164.1
Cotton												
Local						148.8	116.1		146.0	86.6	79.5	140.6
Improved									132.9	110.0	106.4	122.9
Sugarcane					466.0	374.6	512.2	817.1	357.8	358.8	348.3	313.9
Maize												
Local		101.5	72.0			127.6						
Improved									80.7	78.4	95.6	113.7
	Lyallpur				Rahimyar Khan				Jacobabad			
Wheat												
Local					124.4	106.8	118.0	145.9	31.1	33.3	31.0	38.4
Mexi-Pak	109.0	112.8	152.9	217.2	143.0	142.3	162.1	243.6		28.5	26.9	38.5
Rice												
Local	220.0	169.6	147.5			89.4	123.6	67.1	38.5	43.3	34.3	60.0
Basmati	92.7	135.8	116.9	218.1			114.4					
IRRI						160.4			71.8	86.3	61.4	94.8
Cotton												
Local					80.9	74.3	141.7	175.6				
Improved	141.9	130.5	146.2	211.2	135.5	123.5	142.7	307.1				
Sugarcane	380.1	369.4	438.8	388.0		398.8	467.1	488.4				
Maize												
Local	39.4	57.8				62.0	8.0	63.0				
Improved	102.4	89.4	134.1	160.7								

	Larkana				Nawabshah				Hyderabad			
Wheat												
Local	32.7	32.5	36.4	42.1	42.5	43.0	28.1	38.7				
Mexi-Pak					142.2	130.7	128.7	143.8	77.6	101.1	100.8	138.7
Rice												
Local			69.0	51.9					70.2	72.0	60.6	98.7
Basmati	22.5		82.7	54.1								
IRRI	116.3	115.9	106.4	104.0					59.9	68.9	98.8	92.1
Cotton												
Local					103.0	90.4		105.4	22.3	24.4	57.1	127.4
Improved					114.3	123.7	100.8	125.0	99.6	92.0	100.1	116.3
Sugarcane					322.8	961.6	762.7	645.0	583.7	637.9	528.5	497.9
Maize												
Local								59.0	33.0	26.0		
Improved												

Note: Blank spaces indicate data not applicable.

Source: Compiled by the author.

TABLE A.11

Distribution of Average Variable Cost, by Factor Inputs on Crops
(Jhelum, in rupees)

Crop	Seed	Water	Ferti-lizer	Hired Labor	Farm Ma-chinery	Market-ing	Other	Seed	Water	Ferti-lizer	Hired Labor	Farm Ma-chinery	Market-ing	Other
	Under 12.50 Acres							12.50-25.00 Acres						
Wheat														
Local	91.0	32.0	55.0	66.0	102.0		48.0	145.7			250.0	180.6		
	(57.5)	(2.9)	(5.0)	(11.9)	(18.4)		(4.3)	(49.2)			(24.1)	(26.6)		
Mexi-Pak														
Rice														
Local														
Basmati														
IRRI														
Cotton														
Local														
Improved														
Sugarcane														
Maize														
Local								8.0	5.2					
								(60.6)	(39.4)					
Improved														

	25.00-50.00 Acres			Over 50.00 Acres			
Wheat							
Local	304.7		200.0	330.0			120.0
	(82.0)		(18.0)	(73.3)			(26.7)
Mexi-Pak				287.0	70.0	1,193.9	190.0
				(18.5)	(2.3)	(67.0)	(12.2)
Rice							
Local							
Basmati							
IRRI							
Cotton							
Local							
Improved							
Sugarcane							
Maize							
Local	8.0	10.0					
	(44.4)	(55.6)					
Improved							

(continued)

TABLE A.11 (continued)
(Gujrunwala, in rupees)

Crop	Seed	Water	Ferti-lizer	Hired Labor	Farm Ma-chinery	Market-ing	Other	Seed	Water	Ferti-lizer	Hired Labor	Farm Ma-chinery	Market-ing	Other
	Under 12.50 Acres							12.50-25.00 Acres						
Wheat														
Local	22.0	10.3		30.0				150.0	64.6	330.0	125.0			
	(30.4)	(28.2)		(41.4)				(21.4)	(13.9)	(47.0)	(17.8)			
Mexi-Pak	65.0	31.2	110.0	63.8	149.1			243.0	103.3	554.6	245.1	450.0		
	(25.1)	(13.3)	(22.7)	(24.6)	(14.4)			(22.1)	(10.2)	(50.3)	(15.7)	(1.7)		
Rice														
Local	2.0	13.5	55.0	17.5				2.5	14.5	55.0	30.0			
	(2.2)	(19.5)	(59.5)	(18.9)				(2.2)	(22.9)	(48.5)	(26.4)			
Basmati	5.5	28.0	89.4	47.6	78.0			38.3	130.2	485.8	189.1	120.0		
	(3.1)	(17.4)	(46.8)	(25.9)	(6.8)			(4.2)	(21.6)	(57.9)	(15.7)	(0.6)		
IRRI	2.7	22.6	81.8	45.4	115.7			10.3	68.3	196.4	217.1	250.0		
	(1.5)	(13.6)	(44.1)	(24.6)	(16.2)			(2.0)	(18.1)	(50.1)	(27.5)	(2.3)		
Cotton														
Local								12.0	20.8	45.5	25.0			
								(8.1)	(14.0)	(61.2)	(16.8)			
Improved														
Sugarcane	61.9	13.5	55.0	128.1	45.0			191.7	34.0	71.5	130.0			
	(26.6)	(7.2)	(8.9)	(55.0)	(2.4)			(43.7)	(10.6)	(20.9)	(24.7)			
Maize														
Local								8.0	9.7		100.0			
								(6.3)	(15.4)		(78.4)			
Improved														

	25.00-50.00 Acres					Over 50.00 Acres					
Wheat											
Local	175.0	72.0	385.0	340.0		156.3	63.9	366.7	189.5	275.0	
	(18.1)	(11.1)	(39.8)	(31.0)		(16.1)	(13.1)	(28.3)	(28.3)	(14.2)	
Mexi-Pak	553.7	264.6	984.2	367.3	886.7	931.3	398.0	2,186.3	593.0	1,666.6	843.8
	(19.6)	(13.8)	(46.3)	(17.1)	(3.2)	(14.7)	(12.0)	(34.4)	(14.2)	(19.6)	(4.4)
Rice											
Local	20.5	120.0	511.5	291.1		23.0	133.0	550.0	258.5	504.6	206.3
	(2.1)	(24.2)	(51.8)	(22.0)		(1.4)	(14.1)	(34.1)	(24.5)	(20.8)	(5.1)
Basmati	37.1	198.0	551.2	308.0	280.0	46.3	169.6	669.2	347.1	453.4	367.5
	(2.6)	(22.4)	(47.7)	(25.1)	(2.3)	(2.2)	(14.4)	(32.4)	(28.3)	(16.8)	(5.9)
IRRI	23.9	199.6	776.6	342.2	423.3	49.0	359.9	1,562.9	586.5	1,294.4	593.0
	(1.5)	(21.2)	(48.1)	(26.1)	(3.1)	(1.1)	(15.6)	(35.7)	(21.4)	(20.6)	(5.6)
Cotton											
Local	28.0	73.3	440.0	93.6							
	(5.7)	(21.9)	(44.6)	(27.9)							
Improved											
Sugarcane	185.4	36.0	119.2	211.2	100.0	183.3	23.7	99.0	248.8	50.0	27.5
	(28.1)	(8.2)	(16.0)	(45.2)	(2.5)	(24.4)	(5.2)	(11.0)	(54.9)	(3.3)	(1.2)
Maize											
Local											
Improved											

(continued)

TABLE A.11 (continued)
(Sahiwal, in rupees)

Crop	Seed	Water	Ferti- lizer	Hired Labor	Farm Ma- chinery	Market- ing	Other	Seed	Water	Ferti- lizer	Hired Labor	Farm Ma- chinery	Market- ing	Other
	Under 12.50 Acres							12.50-25.00 Acres						
Wheat														
Local	61.9	24.0	102.7	52.0	40.0	52.5	92.0	74.5	38.2	126.4	38.7	64.0		
	(20.7)	(10.9)	(45.6)	(15.1)	(1.1)	(1.5)	(5.1)	(22.6)	(16.4)	(49.5)	(10.6)	(0.9)		
Mexi-Pak	83.4	36.0	175.1	92.8	57.5	49.0	77.8	158.5	71.4	292.6	71.2	187.3	161.9	
	(19.7)	(9.9)	(46.3)	(11.9)	(2.6)	(4.6)	(5.1)	(19.2)	(12.8)	(42.6)	(8.4)	(2.5)	(14.5)	
Rice														
Local	5.0	28.2	110.0	43.3		57.0		7.5	46.8	140.1	34.1		93.0	
	(2.2)	(18.4)	(60.1)	(13.1)		(6.2)		(2.3)	(23.8)	(53.1)	(6.5)		(14.2)	
Basmati	9.5	30.4	119.7	36.2	33.3	53.3	36.8	15.9	62.9	247.7	67.5	137.7	90.6	
	(3.9)	(16.9)	(57.7)	(9.9)	(1.2)	(6.7)	(3.7)	(3.7)	(20.7)	(51.9)	(8.4)	(3.2)	(12.0)	
IRRI	1.0	16.0	55.0	14.0		32.0		5.1	31.2	163.7	62.5		129.8	
	(0.8)	(12.6)	(43.3)	(18.1)		(25.2)		(1.4)	(15.8)	(55.7)	(8.8)		(18.3)	
Cotton														
Local	21.0	41.1	110.0	178.8	46.7	45.0	40.0	15.0	64.8	82.5	167.1	160.0	65.1	
	(5.8)	(14.1)	(7.5)	(49.0)	(9.6)	(3.1)	(11.0)	(3.5)	(23.7)	(6.4)	(47.8)	(6.2)	(12.5)	
Improved	10.7	37.4	124.2	153.0	39.4	45.3	58.4	18.4	86.5	235.3	168.8	167.2	126.3	
	(3.1)	(13.8)	(24.3)	(49.7)	(1.7)	(3.4)	(4.0)	(3.2)	(21.3)	(26.7)	(35.0)	(2.9)	(10.9)	
Sugarcane	184.6	34.9	76.6	111.8	57.5	60.7	30.0	291.3	50.3	156.2	158.6	89.7	79.8	60.0
	(38.2)	(9.6)	(20.5)	(24.0)	(2.3)	(4.1)	(1.2)	(37.8)	(11.0)	(21.6)	(20.8)	(1.3)	(7.2)	(0.3)
Maize														
Local														
Improved	10.7	11.6	59.6	21.9	18.3	37.3	25.6	18.5	22.4	100.5	25.7	45.3	65.2	
	(11.0)	(14.2)	(40.2)	(3.7)	(3.4)	(16.4)	(11.2)	(10.2)	(18.9)	(38.0)	(2.8)	(2.1)	(28.0)	

	25.00-50.00 Acres							Over 50.00 Acres					
Wheat													
Local	100.9	40.5	178.8	69.0	89.0			116.7	48.3	203.9	55.6	121.1	
	(20.8)	(16.0)	(45.5)	(13.7)	(4.1)			(18.3)	(15.1)	(40.4)	(7.2)	(19.0)	
Mexi-Pak	300.6	122.0	565.8	99.8	293.5	377.9		687.3	287.3	1,354.6	154.5	739.7	772.1
	(17.7)	(13.8)	(39.3)	(6.5)	(5.3)	(17.4)		(15.6)	(12.9)	(36.5)	(3.1)	(16.1)	(15.7)
Rice													
Local	9.2	40.7	201.7	40.0	76.0	97.5	10.0	23.1	131.6	517.5	55.0	227.5	226.3
	(2.0)	(16.2)	(43.4)	(5.7)	(10.9)	(21.0)	(0.7)	(1.8)	(19.9)	(39.5)	(4.2)	(17.4)	(17.3)
Basmati	27.0	82.3	312.6	79.5	242.5	155.2	75.0	60.2	182.6	911.1	112.5	343.9	367.9
	(3.5)	(19.8)	(49.7)	(9.2)	(5.0)	(12.7)	(0.2)	(3.1)	(18.4)	(40.8)	(5.3)	(17.0)	(15.4)
IRRI	9.1	49.1	186.3	74.5	105.0	196.8		22.2	120.7	598.7	83.6	216.4	409.7
	(1.8)	(19.0)	(45.9)	(9.6)	(2.8)	(21.0)		(1.4)	(15.8)	(42.1)	(5.4)	(14.0)	(21.2)
Cotton													
Local	22.5	109.0		207.1		97.5		60.0	281.6	1,100.0	300.0	500.0	315.0
	(3.8)	(33.7)		(46.1)		(16.4)		(2.1)	(19.1)	(39.1)	(10.7)	(17.8)	(11.2)
Improved	33.5	138.6	383.0	211.8	285.2	188.0		71.6	300.2	735.9	229.1	621.1	405.1
	(3.2)	(24.4)	(28.1)	(26.7)	(7.4)	(10.2)		(2.6)	(21.2)	(33.0)	(10.2)	(21.7)	(11.2)
Sugarcane	450.0	85.2	186.0	215.0	100.4	127.9		658.8	131.0	364.6	183.9	209.1	129.0
	(38.0)	(11.9)	(18.8)	(20.9)	(2.4)	(8.0)		(38.2)	(12.9)	(24.8)	(8.9)	(11.5)	(3.7)
Maize													
Local													
Improved	26.9	31.6	122.0	58.1	97.0	94.7		51.0	60.8	201.1	75.5	180.6	168.4
	(8.5)	(19.2)	(33.2)	(7.5)	(8.3)	(23.4)		(7.1)	(16.5)	(27.0)	(4.0)	(24.2)	(21.2)

(continued)

TABLE A.11 (continued)
(Lyallpur, in rupees)

Crop	Seed	Water	Fertilizer	Hired Labor	Farm Machinery	Marketing	Other	Seed	Water	Fertilizer	Hired Labor	Farm Machinery	Marketing	Other
	Under 12.50 Acres							12.50-25.00 Acres						
Wheat														
Local														
Mexi-Pak	73.5	37.4	184.0	92.4	50.0			156.8	80.3	355.7	227.4	95.0	114.0	
	(19.0)	(9.8)	(47.6)	(23.3)	(0.3)			(18.7)	(10.3)	(53.4)	(15.2)	(0.6)	(1.8)	
Rice														
Local	3.0	8.0						1.5	8.4	55.0	50.0			
	(27.3)	(72.7)						(1.7)	(9.3)	(61.2)	(27.8)			
Basmati	7.5	12.0	55.0	22.5				4.8	15.7	55.0	171.5			
	(10.8)	(17.3)	(39.6)	(32.4)				(3.5)	(12.7)	(40.5)	(43.3)			
IRRI														
Cotton														
Local														
Improved	15.5	31.5	142.3	209.2	60.0			34.9	74.4	327.0	393.5	125.0	180.0	
	(3.8)	(11.3)	(33.1)	(51.4)	(0.4)			(4.3)	(12.4)	(40.6)	(41.6)	(0.4)	(0.6)	
Sugarcane	129.9	24.7	61.8	117.6	20.0			342.1	67.2	222.4	313.4	35.0	149.3	
	(36.0)	(8.6)	(24.4)	(30.9)	(0.1)			(37.8)	(9.1)	(28.3)	(22.9)	(0.2)	(1.7)	
Maize														
Local	2.9	5.8		25.0				3.3	8.0	55.0				
	(12.3)	(24.3)		(63.4)				(6.9)	(16.7)	(76.4)				
Improved	2.8	6.2	55.0	37.7				5.9	11.5	74.7	124.0		75.0	
	(3.8)	(9.2)	(62.1)	(24.8)				(4.7)	(10.8)	(51.0)	(31.7)		(1.8)	

	25.00-50.00 Acres						Over 50.00 Acres					
Wheat												
Local												
Mexi-Pak	322.1	150.9	948.1	244.7	395.5	312.5	936.0	404.7	2,233.0	414.9	2,112.1	922.9
	(13.7)	(7.5)	(55.6)	(15.1)	(4.6)	(3.5)	(10.4)	(7.4)	(42.5)	(7.9)	(22.8)	(9.0)
Rice												
Local	2.3	24.0	110.0	92.9								
	(1.0)	(10.8)	(24.9)	(63.3)								
Basmati	2.7	26.4	82.5	125.0	27.0		15.2	68.4	372.8	150.1	186.9	60.0
	(1.4)	(15.7)	(42.3)	(34.2)	(6.5)		(1.6)	(13.3)	(38.5)	(25.2)	(19.3)	(2.1)
IRRI												
Cotton												
Local												
Improved	75.2	171.8	706.3	434.1	350.6	478.0	185.9	434.5	1,465.0	647.5	1,713.0	1,578.1
	(4.0)	(11.9)	(37.7)	(38.6)	(4.5)	(3.4)	(2.8)	(11.2)	(25.0)	(18.8)	(24.6)	(17.5)
Sugarcane	1,186.8	167.1	578.2	412.3	215.9	492.5	1,774.1	476.6	1,395.1	636.2	1,056.0	790.4
	(40.5)	(7.5)	(25.8)	(15.2)	(2.1)	(8.8)	(27.0)	(11.1)	(23.7)	(12.8)	(15.6)	(9.9)
Maize												
Local												
Improved	8.6	16.6	123.8	134.1	17.0	76.3	26.8	51.8	191.4	177.4	177.0	127.6
	(3.1)	(7.6)	(44.6)	(41.4)	(0.3)	(3.1)	(3.6)	(10.3)	(24.4)	(28.1)	(22.6)	(11.1)

(continued)

TABLE A.11 (continued)
(Rahimyar Khan, in rupees)

Crop	Seed	Water	Ferti-lizer	Hired Labor	Farm Ma-chinery	Market-ing	Other	Seed	Water	Ferti-lizer	Hired Labor	Farm Ma-chinery	Market-ing	Other
	Under 12.50 Acres							12.50-25.00 Acres						
Wheat														
Local	53.4	73.1	137.9					103.0	93.5	363.0	60.0			
	(17.6)	(42.0)	(40.4)					(19.3)	(22.2)	(56.6)	(1.9)			
Mexi-Pak	52.6	57.7	207.1	175.4	112.2			133.3	147.6	358.7	178.7	467.7		
	(13.0)	(16.2)	(63.1)	(5.0)	(2.6)			(15.1)	(18.0)	(49.0)	(12.8)	(5.1)		
Rice														
Local								75.0	14.4					
								(83.9)	(16.1)					
Basmati														
IRRI								10.0	7.5	55.0				
								(12.5)	(19.0)	(68.6)				
Cotton														
Local	15.7	65.4	74.3			72.5		51.7	153.8	412.5			1.3	
	(8.8)	(53.1)	(26.5)			(11.6)		(9.1)	(42.6)	(48.3)			(0.1)	
Improved	27.1	71.9	265.0	250.0	130.0	151.5		42.5	138.0	455.9	268.7	216.0		
	(5.2)	(22.2)	(56.7)	(4.0)	(2.1)	(9.7)		(5.5)	(27.1)	(52.4)	(12.4)	(2.6)		
Sugarcane	289.3	31.7	109.3	127.5	60.0	1.0		596.5	77.3	193.3	94.2	134.8	19.6	
	(63.7)	(9.0)	(23.1)	(3.3)	(0.9)			(58.2)	(13.0)	(22.6)	(4.7)	(1.1)	(0.5)	
Maize														
Local								7.0		55.0				
								(11.3)		(88.7)				
Improved														

	25.00-50.00 Acres						Over 50.00 Acres					
Wheat												
Local	155.3	148.1	427.5	242.7	32.0		200.0	300.0	550.0	385.0	24.0	
	(16.5)	(18.3)	(45.3)	(18.7)	(1.3)		(13.7)	(20.6)	(37.7)	(26.4)	(1.6)	
Mexi-Pak	184.8	188.2	583.1	299.1	413.4		547.8	532.7	2,736.6	1,667.4	2,406.6	70.0
	(12.9)	(14.9)	(44.7)	(14.4)	(13.1)		(8.1)	(8.2)	(52.5)	(17.1)	(14.1)	(0.1)
Rice												
Local	54.3	31.9	220.0	21.7	72.0		150.0	242.4		100.0	72.0	
	(26.3)	(16.0)	(35.5)	(10.5)	(11.6)		(32.0)	(48.7)		(14.2)	(5.1)	
Basmati	30.0	28.8	110.0	60.0								
	(13.1)	(12.6)	(48.1)	(26.2)								
IRRI												
Cotton												
Local	120.0	409.8	1,815.0	300.0	72.0		298.7	682.6	3,006.7	753.5	1,853.2	27.0
	(3.9)	(26.0)	(58.2)	(9.6)	(2.3)		(5.3)	(20.2)	(53.5)	(8.9)	(11.8)	(0.2)
Improved	65.6	227.9	604.9	402.2	320.0		303.7	1,014.4	6,831.8	1,648.0	4,093.2	80.0
	(4.5)	(21.7)	(48.7)	(17.5)	(7.5)		(2.4)	(10.5)	(69.3)	(5.9)	(11.9)	(0.1)
Sugarcane	999.9	150.7	437.1	208.2	231.0		2,239.6	397.1	2,768.7	351.2	357.2	38.0
	(52.1)	(10.8)	(26.6)	(6.9)	(3.6)		(35.9)	(9.1)	(50.0)	(3.3)	(1.8)	(0.1)
Maize												
Local		4.0					3.5	4.0			24.0	
		(100.0)					(11.1)	(12.7)			(76.2)	
Improved												

(continued)

TABLE A. 11 (continued)
(Jacobabad, in rupees)

Crop	Seed	Water	Ferti-lizer	Hired Labor	Farm Ma-chinery	Market-ing	Other	Seed	Water	Ferti-lizer	Hired Labor	Farm Ma-chinery	Market-ing	Other
	Under 12.50 Acres							12.50-25.00 Acres						
Wheat														
Local	198.0					36.7		250.0					27.4	
	(87.8)					(12.2)		(90.1)					(9.9)	
Mexi-Pak								204.0					20.0	
								(95.3)					(4.7)	
Rice														
Local	208.8	40.5	128.0			84.7		414.0		335.4		690.0	170.9	64.0
	(59.7)	(2.8)	(13.3)			(24.2)		(51.0)		(18.8)		(7.7)	(21.0)	(1.4)
Basmati														
IRRI	50.0		58.0			35.6		550.0		986.0		220.0	142.0	
	(34.8)		(40.4)			(24.8)		(29.0)		(51.9)		(11.6)	(7.5)	
Cotton														
Local														
Improved														
Sugarcane														
Maize														
Local														
Improved														

	25.00-50.00 Acres						Over 50.00 Acres					
Wheat												
Local	420.0				14.0		572.8	85.7		44.8	240.0	45.0
	(96.8)				(3.2)		(76.1)	(8.0)		(3.6)	(6.4)	(6.0)
Mexi-Pak	240.3				8.8		620.7	174.3		108.0	600.0	95.4
	(97.8)				(2.2)		(67.9)	(11.3)		(4.2)	(7.3)	(9.3)
Rice												
Local	639.9		386.3		254.1		1,653.8	653.0	1,351.0	249.7	1,697.4	514.2
	(57.3)		(22.0)		(20.7)		(40.6)	(5.7)	(30.1)	(3.3)	(7.7)	(12.6)
Basmati												
IRRI	438.3		1,073.0		149.5	80.0	1,262.3	494.7	1,111.0	395.3	1,476.5	513.0
	(33.0)		(53.8)		(11.2)	(2.0)	(26.9)	(5.0)	(23.7)	(2.5)	(31.1)	(10.9)
Cotton												
Local												
Improved												
Sugarcane												
Maize												
Local												
Improved												

(continued)

TABLE A.11 (continued)
(Larkana, in rupees)

Crop	Seed	Water	Fertilizer	Hired Labor	Farm Machinery	Marketing	Other	Seed	Water	Fertilizer	Hired Labor	Farm Machinery	Marketing	Other
	Under 12.50 Acres							12.50-25.00 Acres						
Wheat														
Local	280.7					37.8		343.6					25.4	
	(95.7)					(4.3)		(95.3)					(4.7)	
Mexi-Pak														
Rice														
Local														
Basmati	45.0													
	(100.0)													
IRRI	100.4		649.9			157.1		245.9		1,608.8			380.3	
	(11.1)		(71.6)			(17.3)		(11.0)		(72.0)			(17.0)	
Cotton														
Local														
Improved														
Sugarcane														
Maize														
Local														
Improved														

	25.00-50.00 Acres						Over 50.00 Acres					
Wheat												
Local	567.2	150.0	275.0			33.1	848.5		473.0			60.3
	(88.1)	(3.2)	(3.6)			(5.1)	(76.7)		(17.8)			(5.5)
Mexi-Pak												
Rice												
Local	50.0		220.0			6.0	47.2		73.3			27.0
	(18.1)		(79.7)			(2.2)	(34.1)		(52.9)			(13.0)
Basmati	62.5		165.0			20.5	102.5		110.0			62.0
	(25.2)		(66.5)			(8.3)	(42.1)		(45.2)			(12.7)
IRRI	473.7		2,713.3			589.6	779.8		4,301.7		3,510.0	724.9
	(12.5)		(71.8)			(15.6)	(13.0)		(72.0)		(4.9)	(10.1)
Cotton												
Local												
Improved												
Sugarcane												
Maize												
Local												
Improved												

(continued)

TABLE A.11 (continued)
(Nawabshah, in rupees)

Crop	Seed	Water	Fertilizer	Hired Labor	Farm Machinery	Marketing	Other	Seed	Water	Fertilizer	Hired Labor	Farm Machinery	Marketing	Other
	Under 12.50 Acres							12.50-25.00 Acres						
Wheat														
Local	272.4		168.0			36.9		300.5		112.0			33.3	
	(82.5)		(6.4)			(11.2)		(83.1)		(7.7)			(9.2)	
Mexi-Pak	165.6		641.5			124.0		380.0		1,428.0			219.8	
	(17.8)		(68.9)			(13.3)		(18.7)		(70.4)			(10.8)	
Rice														
Local														
Basmati														
IRRI														
Cotton														
Local	81.0		546.0			60.1		80.0		466.7			25.7	
	(11.8)		(79.5)			(8.8)		(14.0)		(75.9)			(4.5)	
Improved	101.3		609.0			78.4		130.0		888.8			151.6	
	(12.8)		(77.2)			(9.9)		(11.1)		(75.9)			(13.0)	
Sugarcane	542.7		373.3			304.3		2,480.0		1,372.8			8,936.3	
	(44.5)		(30.6)			(24.9)		(19.4)		(10.7)			(69.9)	
Maize														
Local														
Improved														

	25.00-50.00 Acres						Over 50.00 Acres					
Wheat												
Local	462.9					71.8	760.5	192.0	1,100.0			91.1
	(88.6)					(11.4)	(72.5)	(5.7)	(13.1)			(8.7)
Mexi-Pak	525.0		1,891.0			287.1	875.0	83.1	2,856.0		1,680.0	549.2
	(19.4)		(70.0)			(10.6)	(18.7)	(1.5)	(60.9)		(7.2)	(11.7)
Rice												
Local												
Basmati												
IRRI												
Cotton												
Local							375.0		2,035.0			225.0
							(14.2)		(77.2)			(8.5)
Improved	231.8		1,639.4			157.8	482.5	172.9	3,126.7		2,520.0	276.1
	(11.4)		(80.8)			(7.8)	(11.6)	(1.7)	(75.1)		(5.0)	(6.6)
Sugarcane	4,049.8		2,470.7			2,058.4	6,164.4	205.8	4,017.4		3,500.0	4,423.4
	(21.8)		(13.3)			(64.9)	(24.7)	(0.4)	(16.1)		(1.2)	(57.7)
Maize												
Local							85.0	60.0	275.0			110.0
Improved							(14.4)	(20.3)	(46.6)			(18.6)

(continued)

TABLE A.11 (continued)
(Hyderabad, in rupees)

Crop	Seed	Water	Ferti-lizer	Hired Labor	Farm Ma-chinery	Market-ing	Other	Seed	Water	Ferti-lizer	Hired Labor	Farm Ma-chinery	Market-ing	Other
	Under 12.50 Acres							12.50-25.00 Acres						
Wheat														
Local														
Mexi-Pak	88.8	8.6	171.8	47.6	90.0	30.0	150.0	172.8	16.0	356.7	68.3		66.1	25.0
	(30.2)	(1.1)	(48.2)	(7.7)	(1.6)	(8.6)	(2.7)	(27.7)	(0.5)	(54.7)	(7.0)		(9.9)	(0.2)
Rice														
Local	14.0		84.0	45.0		15.0		31.5		126.0	62.5		14.0	
	(8.9)		(53.2)	(28.5)		(9.5)		(13.5)		(53.8)	(26.7)		(6.0)	
Basmati														
IRRI	77.9		209.0	59.3		60.6		144.9		392.4	106.1		107.8	
	(22.4)		(48.2)	(12.0)		(17.5)		(19.6)		(53.0)	(12.9)		(14.6)	
Cotton														
Local	15.0		18.5	26.0		16.0		60.0			60.0		75.0	
	(33.6)		(10.4)	(29.1)		(26.9)		(30.8)			(30.8)		(38.5)	
Improved	44.5		213.8	46.7	90.0	45.3	50.0	57.6	16.0	296.9	127.8		102.4	
	(15.4)		(59.2)	(4.8)	(3.1)	(15.7)	(1.7)	(10.4)	(0.8)	(55.3)	(14.9)		(18.5)	
Sugarcane	535.0		252.2	202.5		531.5		726.3		165.0	45.7		1,003.1	
	(37.6)		(18.0)	(7.1)		(37.3)		(37.9)		(7.5)	(2.1)		(52.4)	
Maize														
Local	15.0					18.0		15.0					11.0	
	(45.5)					(54.5)		(57.7)					(42.3)	
Improved														

	25.00-50.00 Acres							Over 50.00 Acres					
Wheat													
Local													
Mexi-Pak	264.5	12.4	706.4	204.3		101.2	100.0	745.1	56.1	1,724.9	534.2	970.0	300.4
	(25.4)	(0.3)	(54.2)	(9.7)		(9.7)	(0.6)	(19.4)	(0.6)	(52.8)	(11.5)	(7.9)	(7.8)
Rice													
Local	56.0		252.0	40.0		35.5		136.7		630.0	241.7	600.0	106.7
	(15.4)		(69.3)	(5.5)		(9.8)		(10.4)		(47.9)	(18.4)	(15.2)	(8.1)
Basmati													
IRRI	275.0		1,095.0	374.3		197.8		417.5		1,142.3	489.1	1,200.0	327.0
	(15.5)		(61.6)	(11.8)		(11.1)		(15.2)		(52.0)	(12.1)	(8.7)	(11.9)
Cotton													
Local	30.0		168.0	80.0		34.5		108.8		1,134.0	422.9	240.0	141.7
	(13.1)		(36.8)	(35.0)		(15.1)		(6.5)		(67.6)	(12.7)	(4.8)	(8.4)
Improved	57.8	5.3	694.7	302.1		98.0	50.0	331.5	38.8	2,120.0	545.2	1,106.3	458.3
	(6.7)	(0.5)	(62.3)	(18.6)		(11.3)	(0.6)	(8.5)	(0.9)	(54.5)	(11.6)	(12.6)	(11.8)
Sugarcane	1,415.5	1.2	378.5	271.4		1,215.3		1,817.1	22.5	506.4	256.0	397.5	1,083.8
	(46.7)		(12.5)	(4.2)		(36.5)		(50.3)	(0.1)	(15.2)	(5.5)	(3.1)	(25.7)
Maize													
Local													
Improved													

Notes: Figures given in parentheses are the percentage shares of each input in the total input costs for each crop. Blank spaces indicate data not applicable.

Source: Compiled by the author.

TABLE A.12

Distribution of Average Variable Cost per Acre, by Factor Inputs on Crops
(Jhelum, in rupees)

Crop	Seed	Water	Ferti-lizer	Hired Labor	Farm Ma-chinery	Market-ing	Other	Seed	Water	Ferti-lizer	Hired Labor	Farm Ma-chinery	Market-ing	Other
	Under 12.50 Acres							12.50-25.00 Acres						
Wheat														
Local	23.6	8.3	14.3	17.1	26.4	0.0	12.4	19.4	0.0	0.0	33.3	24.1	0.0	0.0
Mexi-Pak														
Rice														
Local														
Basmati														
IRRI														
Cotton														
Local														
Improved														
Sugarcane														
Maize														
Local								61.5	40.0	0.0	0.0	0.0	0.0	0.0
Improved														
	25.00-50.00 Acres							Over 50.00 Acres						
Wheat														
Local	18.3	0.0	0.0	12.0	0.0	0.0	0.0	16.5	0.0	0.0	0.0	6.0	0.0	0.0
Mexi-Pak								14.4	0.0	3.5	59.7	9.5	0.0	0.0
Rice														
Local														
Basmati														
IRRI														
Cotton														
Local														
Improved														
Sugarcane														
Maize														
Local	32.0	40.0	0.0	0.0	0.0	0.0	0.0							
Improved														

(Gujranwala, in rupees)

	Under 12.50 Acres							12.50-25.00 Acres						
Wheat														
Local	22.0	10.3	0.0	30.0	0.0	0.0	0.0	25.0	10.8	55.0	20.8	0.0	0.0	0.0
Mexi-Pak	21.3	10.2	36.1	20.9	48.9	0.0	0.0	24.8	10.5	56.4	24.9	45.8	0.0	0.0
Rice														
Local	2.0	13.5	55.0	17.5	0.0	0.0	0.0	2.5	14.5	55.0	30.0	0.0	0.0	0.0
Basmati	2.7	13.7	43.8	23.3	38.2	0.0	0.0	4.3	14.7	54.8	21.3	13.5	0.0	0.0
IRRI	1.5	12.3	44.3	24.7	62.9	0.0	0.0	2.1	13.8	39.7	43.9	50.5	0.0	0.0
Cotton														
Local								12.0	20.8	45.5	25.0	0.0	0.0	0.0
Improved														
Sugarcane	123.8	27.0	110.0	256.3	90.0	0.0	0.0	163.8	29.0	61.1	111.1	0.0	0.0	0.0
Maize														
Local								8.0	9.7	0.0	100.0	0.0	0.0	0.0
Improved														
	25.00-50.00 Acres							Over 50.00 Acres						
Wheat														
Local	25.0	10.3	55.0	48.6	0.0	0.0	0.0	25.0	10.2	58.7	30.3	44.0	0.0	0.0
Mexi-Pak	23.1	11.1	41.1	15.4	37.1	0.0	0.0	23.0	9.8	53.9	14.6	41.1	20.8	0.0
Rice														
Local	2.3	13.2	56.2	32.0	0.0	0.0	0.0	2.3	13.3	55.0	25.9	50.5	20.6	0.0
Basmati	3.1	16.5	45.9	25.7	23.3	0.0	0.0	3.8	13.9	55.0	28.5	37.3	30.2	0.0
IRRI	1.7	13.8	53.8	23.7	29.3	0.0	0.0	1.7	12.7	55.0	20.6	45.6	20.9	0.0
Cotton														
Local	6.6	17.3	103.5	22.0	0.0	0.0	0.0							
Improved														
Sugarcane	143.7	27.9	92.4	163.7	77.5	0.0	0.0	199.3	25.8	107.6	270.4	54.4	29.9	0.0
Maize														
Local														
Improved														

(continued)

TABLE A. 12 (continued)
(Sahiwal, in rupees)

Crop	Seed	Water	Ferti-lizer	Hired Labor	Farm Ma-chinery	Market-ing	Other	Seed	Water	Ferti-lizer	Hired Labor	Farm Ma-chinery	Market-ing	Other
	Under 12.50 Acres							12.50-25.00 Acres						
Wheat														
Local	24.8	9.6	41.1	20.8	16.0	21.0	36.9	24.0	12.3	40.8	12.5	20.7	0.0	0.0
Mexi-Pak	23.9	10.3	50.2	26.6	16.5	14.0	22.3	24.5	11.1	45.3	11.0	29.0	25.1	0.0
Rice														
Local	2.5	14.1	55.0	21.7	0.0	28.5	0.0	2.5	15.6	46.7	11.4	0.0	31.0	0.0
Basmati	4.4	14.6	57.3	17.3	16.0	25.5	17.6	4.4	17.4	68.4	18.7	38.0	25.0	0.0
IRRI	1.0	16.0	55.0	14.8	0.0	32.0	0.0	2.3	13.9	72.8	27.8	0.0	57.7	0.0
Cotton														
Local	8.4	16.5	44.0	71.5	18.7	18.0	16.0	3.0	13.0	16.5	33.4	32.0	13.0	0.0
Improved	4.1	14.4	47.8	58.8	15.1	17.4	22.5	3.5	16.4	44.6	32.0	31.7	23.9	0.0
Sugarcane	136.8	25.8	56.7	82.8	42.6	45.0	22.2	135.5	23.4	72.7	73.7	41.7	37.1	27.9
Maize														
Local														
Improved	8.9	9.6	49.2	18.1	15.2	30.9	21.1	8.0	9.7	43.5	11.1	19.6	28.2	0.0
	25.00-50.00 Acres							Over 50.00 Acres						
Wheat														
Local	24.8	10.0	43.9	16.9	21.9	0.0	0.0	24.7	10.2	43.1	11.8	25.6	0.0	0.0
Mexi-Pak	24.4	9.9	45.9	8.1	23.8	30.7	0.0	24.1	10.1	47.5	5.4	25.9	27.1	0.0
Rice														
Local	2.5	11.1	55.0	10.9	20.7	26.6	2.7	3.0	14.1	55.0	15.0	25.0	15.8	0.0
Basmati	4.4	13.5	51.2	13.0	39.8	25.4	12.3	4.3	13.1	65.6	8.1	24.7	26.5	0.0
IRRI	2.5	13.4	50.8	20.3	28.6	53.6	0.0	2.4	12.8	63.7	8.9	23.0	43.6	0.0
Cotton														
Local	3.0	14.5	0.0	27.6	0.0	13.0	0.0	3.2	13.5	33.1	10.3	27.9	18.2	0.0
Improved	3.4	14.1	38.9	21.5	28.9	19.1	0.0	3.0	14.1	55.0	15.0	25.0	15.8	0.0
Sugarcane	132.4	25.1	54.7	63.2	29.5	37.6	0.0	119.8	23.8	66.3	33.4	38.0	23.5	0.0
Maize														
Local														
Improved	8.1	9.6	36.8	17.5	29.2	28.5	0.0	8.0	9.6	31.6	11.9	28.4	26.5	0.0

(Lyallpur, in rupees)

	Under 12.50 Acres							12.50-25.00 Acres						
Wheat														
Local														
Mexi-Pak	20.7	10.5	51.8	26.0	14.1	0.0	0.0	21.0	10.8	47.7	30.5	12.8	15.3	0.0
Rice														
Local	60.0	16.0	0.0	0.0	0.0	0.0	0.0	2.8	15.9	103.8	94.3	0.0	0.0	0.0
Basmati	10.0	16.0	73.3	30.0	0.0	0.0	0.0	4.8	15.7	55.0	171.5	0.0	0.0	0.0
IRRI														
Cotton														
Local														
Improved	5.4	11.0	49.6	72.9	20.9	0.0	0.0	5.7	12.1	53.0	63.8	20.3	29.2	0.0
Sugarcane	136.7	26.0	65.1	123.8	21.1	0.0	0.0	139.6	27.4	90.8	127.9	14.3	60.9	0.0
Maize														
Local	4.8	9.6	0.0	41.7	0.0	0.0	0.0	4.0	9.6	66.3	0.0	0.0	0.0	0.0
Improved	3.9	8.7	77.5	47.4	0.0	0.0	0.0	4.2	8.3	53.8	89.2	0.0	0.0	0.0
	25.00-50.00 Acres							Over 50.00 Acres						
Wheat														
Local														
Mexi-Pak	20.9	9.8	61.5	15.9	25.7	20.3	0.0	22.6	9.8	53.9	10.0	50.9	22.3	0.0
Rice														
Local	1.5	16.0	73.3	61.9	0.0	0.0	0.0							
Basmati	1.6	15.8	49.4	74.9	16.2	0.0	0.0	3.4	15.4	84.0	33.8	42.1	13.5	0.0
IRRI														
Cotton														
Local														
Improved	5.9	13.4	55.1	33.8	27.3	37.3	0.0	5.8	13.6	45.9	20.3	53.6	49.4	0.0
Sugarcane	177.7	25.0	86.6	61.7	32.3	73.7	0.0	104.8	28.2	82.4	37.6	62.4	46.7	0.0
Maize														
Local														
Improved	4.2	8.0	59.8	64.8	8.2	36.8	0.0	5.7	11.0	40.8	37.8	37.7	27.2	0.0

(continued)

TABLE A.12 (continued)
(Rahimyar Khan, in rupees)

Crop	Seed	Water	Ferti-lizer	Hired Labor	Farm Ma-chinery	Market-ing	Other	Seed	Water	Ferti-lizer	Hired Labor	Farm Ma-chinery	Market-ing	Other
	Under 12.50 Acres							12.50-25.00 Acres						
Wheat														
Local	21.9	29.9	56.5	0.0	0.0	0.0	0.0	20.6	18.7	72.6	12.0	0.0	0.0	0.0
Mexi-Pak	18.7	20.5	73.5	62.2	39.8	0.0	0.0	21.5	23.8	57.9	28.8	75.4	0.0	0.0
Rice														
Local								75.0	14.4	0.0	0.0	0.0	0.0	0.0
Basmati														
IRRI								20.0	14.9	110.0	0.0	0.0	0.0	0.0
Cotton														
Local	7.1	29.6	33.6	0.0	0.0	32.8	0.0	6.7	20.1	53.8	0.0	0.0	0.1	0.0
Improved	7.1	18.8	69.2	65.3	33.9	39.6	0.0	6.8	22.2	73.4	43.3	34.8	0.0	0.0
Sugarcane	301.4	33.1	113.9	132.8	62.5	1.0	0.0	232.1	30.1	75.2	36.6	52.4	7.6	0.0
Maize														
Local								7.0	0.0	55.0	0.0	0.0	0.0	0.0
Improved														
	25.00-50.00 Acres							Over 50.00 Acres						
Wheat														
Local	19.4	18.5	53.4	30.3	4.0	0.0	0.0	20.0	30.0	55.0	38.5	2.4	0.0	0.0
Mexi-Pak	20.9	21.3	66.0	33.9	46.8	0.0	0.0	19.7	19.2	98.5	60.0	86.7	2.5	0.0
Rice														
Local	32.5	19.1	131.7	13.0	43.1	0.0	0.0	21.4	34.6	0.0	14.3	10.3	0.0	0.0
Basmati	15.0	14.4	55.0	30.0	0.0	0.0	0.0							
IRRI														
Cotton														
Local	5.5	18.6	82.5	13.6	3.3	0.0	0.0	9.3	21.3	94.0	23.6	57.9	0.8	0.0
Improved	6.5	22.4	59.5	39.6	31.5	0.0	0.0	7.3	24.4	164.4	39.7	98.5	1.9	0.0
Sugarcane	243.5	37.0	107.4	51.2	56.8	0.0	0.0	175.1	31.0	216.5	27.5	27.9	3.0	0.0
Maize														
Local	0.0	8.0	0.0	0.0	0.0	0.0	0.0	7.0	8.0	0.0	0.0	48.0	0.0	0.0
Improved														

(Jacobabad, in rupees)

	Under 12.50 Acres							12.50-25.00 Acres						
Wheat														
Local	27.3	0.0	0.0	0.0	0.0	5.1	0.0	30.0	0.0	0.0	0.0	0.0	3.3	0.0
Mexi-Pak								27.2	0.0	0.0	0.0	0.0	2.7	0.0
Rice														
Local	23.0	4.5	14.1	0.0	0.0	9.3	0.0	22.1	0.0	17.9	0.0	36.8	9.1	3.4
Basmati														
IRRI	25.0	0.0	29.0	0.0	0.0	17.8	0.0	25.0	0.0	44.8	0.0	10.0	6.5	0.0
Cotton														
Local														
Improved														
Sugarcane														
Maize														
Local														
Improved														
	25.00-50.00 Acres							Over 50.00 Acres						
Wheat														
Local	30.0	0.0	0.0	0.0	0.0	1.0	0.0	29.2	4.4	0.0	2.3	12.3	2.3	0.0
Mexi-Pak	26.3	0.0	0.0	0.0	0.0	1.0	0.0	26.2	7.4	0.0	4.6	25.3	4.0	0.0
Rice														
Local	19.7	0.0	11.9	0.0	0.0	7.8	0.0	24.4	9.6	19.9	3.7	25.0	7.6	0.0
Basmati														
IRRI	20.2	0.0	49.5	0.0	0.0	6.9	3.7	25.5	10.0	22.4	8.0	29.8	10.4	0.0
Cotton														
Local														
Improved														
Sugarcane														
Maize														
Local														
Improved														

(continued)

TABLE A.12 (continued)
(Larkana, in rupees)

Crop	Seed	Water	Ferti-lizer	Hired Labor	Farm Ma-chinery	Market-ing	Other	Seed	Water	Ferti-lizer	Hired Labor	Farm Ma-chinery	Market-ing	Other
	Under 12.50 Acres							12.50-25.00 Acres						
Wheat														
Local	31.3	0.0	0.0	0.0	0.0	4.2	0.0	31.0	0.0	0.0	0.0	0.0	2.3	0.0
Mexi-Pak														
Rice														
Local														
Basmati	22.5	0.0	0.0	0.0	0.0	0.0	0.0							
IRRI	12.9	0.0	83.3	0.0	0.0	20.1	0.0	12.8	0.0	83.4	0.0	0.0	19.7	0.0
Cotton														
Local														
Improved														
Sugarcane														
Maize														
Local														
Improved														
	25.00-50.00 Acres							Over 50.00 Acres						
Wheat														
Local	32.1	8.5	15.6	0.0	0.0	1.9	0.0	32.3	0.0	18.0	0.0	0.0	2.3	0.0
Mexi-Pak														
Rice														
Local	12.5	0.0	55.0	0.0	0.0	1.5	0.0	17.7	0.0	27.5	0.0	0.0	10.1	0.0
Basmati	20.8	0.0	55.0	0.0	0.0	6.8	0.0	22.8	0.0	24.4	0.0	0.0	13.8	0.0
IRRI	13.3	0.0	76.4	0.0	0.0	16.6	0.0	13.6	0.0	74.8	0.0	61.0	12.6	0.0
Cotton														
Local														
Improved														
Sugarcane														
Maize														
Local														
Improved														

(Nawabshah, in rupees)

	Under 12.50 Acres							12.50-25.00 Acres						
Wheat														
Local	35.0	0.0	21.6	0.0	0.0	4.7	0.0	35.7	0.0	13.0	0.0	0.0	4.0	0.0
Mexi-Pak	25.3	0.0	97.9	0.0	0.0	18.9	0.0	24.5	0.0	92.1	0.0	0.0	14.2	0.0
Rice														
Local														
Basmati														
IRRI														
Cotton														
Local	12.1	0.0	81.9	0.0	0.0	9.0	0.0	12.6	0.0	73.7	0.0	0.0	4.1	0.0
Improved	14.7	0.0	88.3	0.0	0.0	11.4	0.0	13.7	0.0	94.0	0.0	0.0	16.0	0.0
Sugarcane	143.6	0.0	98.8	0.0	0.0	80.5	0.0	186.5	0.0	103.2	0.0	0.0	671.9	0.0
Maize														
Local														
Improved														
	25.00-50.00 Acres							Over 50.00 Acres						
Wheat														
Local	24.9	0.0	0.0	0.0	0.0	3.9	0.0	28.0	7.1	40.6	0.0	0.0	3.4	0.0
Mexi-Pak	25.0	0.0	90.1	0.0	0.0	13.7	0.0	26.8	2.6	87.6	0.0	51.5	16.9	0.0
Rice														
Local														
Basmati														
IRRI														
Cotton														
Local								15.0	0.0	81.4	0.0	0.0	9.0	0.0
Improved	11.5	0.0	81.5	0.0	0.0	7.8	0.0	14.5	5.2	93.8	0.0	75.6	8.3	0.0
Sugarcane	166.3	0.0	101.4	0.0	0.0	495.0	0.0	159.1	5.3	103.7	0.0	90.3	372.2	0.0
Maize														
Local								8.5	6.0	27.5	0.0	0.0	11.0	0.0
Improved														

(continued)

TABLE A.12 (continued)
(Hyderabad, in rupees)

Crop	Seed	Water	Ferti- lizer	Hired Labor	Farm Ma- chinery	Market- ing	Other	Seed	Water	Ferti- lizer	Hired Labor	Farm Ma- chinery	Market- ing	Other
	Under 12.50 Acres							12.50-25.00 Acres						
Wheat														
Local														
Mexi-Pak	23.4	2.3	45.3	12.6	23.8	7.9	39.6	28.0	2.6	57.7	11.1	0.0	10.7	4.1
Rice														
Local	6.2	0.0	37.3	20.0	0.0	6.7	0.0	9.7	0.0	38.8	19.2	0.0	4.3	0.0
Basmati														
IRRI	13.4	0.0	36.0	10.2	0.0	10.5	0.0	13.5	0.0	36.5	9.9	0.0	10.0	0.0
Cotton														
Local	7.5	0.0	9.3	13.0	0.0	8.0	0.0	7.5	0.0	0.0	7.5	0.0	9.4	0.0
Improved	15.3	0.0	73.7	16.1	31.0	15.6	17.2	9.6	2.7	49.5	21.3	0.0	17.1	0.0
Sugarcane	219.3	0.0	103.4	83.0	0.0	217.8	0.0	242.1	0.0	55.0	15.2	0.0	334.4	0.0
Maize														
Local	15.0	0.0	0.0	0.0	0.0	18.0	0.0	15.0	0.0	0.0	0.0	0.0	11.0	0.0
Improved														
	25.00-50.00 Acres							Over 50.00 Acres						
Wheat														
Local														
Mexi-Pak	25.7	1.2	68.5	19.8	0.0	9.8	9.7	26.9	2.0	62.3	19.3	35.0	10.9	0.0
Rice														
Local	9.3	0.0	42.0	6.7	0.0	5.9	0.0	10.3	0.0	47.3	18.1	45.0	8.0	0.0
Basmati														
IRRI	15.3	0.0	60.8	21.0	0.0	11.0	0.0	14.0	0.0	38.3	16.4	40.3	11.0	0.0
Cotton														
Local	7.5	0.0	42.0	20.0	0.0	8.6	0.0	8.3	0.0	86.1	32.1	18.2	10.8	0.0
Improved	6.7	0.6	80.1	34.9	0.0	11.3	5.8	9.9	1.2	63.4	16.3	33.1	13.7	0.0
Sugarcane	247.0	0.2	66.1	47.4	0.0	212.1	0.0	250.6	3.1	69.9	35.3	54.9	149.5	0.0
Maize														
Local														
Improved														

Note: Blank spaces indicate data not applicable.

Source: Compiled by the author.

TABLE A.13

Distribution of Average Net Farm Income, by Crop
(rupees)

Crop	Under 12.50 Acres Amount	(%)	12.50-25.00 Acres Amount	(%)	25.00-50.00 Acres Amount	(%)	Over 50.00 Acres Amount	(%)
				Jhelum				
Wheat								
Local	266.5	28.1	326.9	11.5	1,414.7	40.1	3,350.0	44.1
Mexi-Pak							442.5	11.6
Rice								
Local								
Basmati								
IRRI								
Cotton								
Local								
Improved								
Sugarcane								
Maize								
Local			56.8	0.3	101.0	0.9		
Improved								
				Gujranwala				
Wheat								
Local	117.6	0.1	497.6	0.4	1,027.2	0.3	1,301.3	1.2
Mexi-Pak	652.3	20.1	2,818.5	24.9	7,013.9	33.5	13,233.4	37.2
Rice								
Local	445.0	1.0	286.5	0.2	2,556.3	4.2	3,479.6	4.1
Basmati	1,214.0	34.7	3,671.2	29.7	6,042.5	25.9	7,319.3	20.6
IRRI	688.3	21.2	1,345.2	10.4	4,424.3	18.2	9,626.8	27.1
Cotton								
Local			61.2	0.0	1,381.5	0.4		
Improved								
Sugarcane	712.0	6.3	1,276.7	2.8	1,199.7	2.4	467.4	0.6
Maize								
Local			32.4	0.0				
Improved								
				Sahiwal				
Wheat								
Local	691.3	2.3	833.0	2.1	1,069.2	1.8	1,215.6	0.5
Mexi-Pak	1,095.4	13.1	2,607.3	14.5	5,328.6	15.5	12,451.4	16.6
Rice								
Local	613.8	0.7	895.9	0.9	1,019.0	0.2	2,393.1	0.3
Basmati	1,235.9	11.7	2,279.3	10.7	3,753.2	9.7	14,253.8	17.0
IRRI	473.0	0.1	962.0	0.4	1,667.6	1.6	4,637.3	0.7
Cotton								
Local	1,537.4	1.7	3,264.8	2.3	5,309.0	1.3	16,089.0	0.5
Improved	2,967.4	34.7	7,062.2	34.1	12,955.3	34.4	28,913.3	37.8
Sugarcane	1,608.0	17.9	2,378.1	13.2	4,268.7	11.6	6,188.8	6.8
Maize								
Local	75.0	0.0						
Improved	267.7	1.6	509.3	2.2	619.5	1.5	1,250.4	1.2

(continued)

TABLE A.13 (continued)

Crop	Under 12.50 Acres		12.50-25.00 Acres		25.00-50.00 Acres		Over 50.00 Acres	
	Amount	(%)	Amount	(%)	Amount	(%)	Amount	(%)
				Lyallpur				
Wheat								
Local								
Mexi-Pak	1,232.8	20.1	2,743.4	19.8	6,311.9	17.7	15,711.1	15.5
Rice								
Local	97.0	0.0	99.1	0.0	268.8	0.0		
Basmati	272.5	0.2	441.8	0.7	937.8	0.4	1,784.5	0.5
IRRI								
Cotton								
Local								
Improved	1,826.0	29.8	4,748.2	34.3	12,695.2	35.6	40,625.8	37.8
Sugarcane	1,020.1	16.7	2,954.5	21.4	10,991.8	30.8	33,453.2	33.0
Maize								
Local	75.9	0.2	145.3	0.1				
Improved	298.4	3.6	706.4	4.4	1,105.5	2.9	2,045.8	1.5
				Rahimyar Khan				
Wheat								
Local	193.9	1.9	566.0	1.6	613.1	1.4	1,353.5	0.1
Mexi-Pak	564.1	9.4	1,192.5	8.8	1,197.0	5.1	3,568.0	6.1
Rice								
Local			810.6	0.4	423.6	0.4	270.5	0.1
Basmati					2,021.2	0.6		
IRRI			405.8	0.2				
Cotton								
Local	1,488.3	11.5	3,643.8	5.3	14,883.4	4.3	15,765.1	6.2
Improved	2,725.5	36.2	5,050.9	42.1	7,684.7	41.9	29,718.6	42.8
Sugarcane	1,110.1	17.2	3,217.4	23.7	5,805.0	31.7	16,854.3	30.9
Maize								
Local	96.0	0.1	130.0	0.1	92.0	0.0	32.5	0.0
Improved								
				Jacobabad				
Wheat								
Local	1,337.0	10.3	1,557.2	4.1	1,930.0	1.8	4,595.2	4.0
Mexi-Pak			1,091.0	3.8	1,316.9	4.9	4,598.3	7.1
Rice								
Local	3,156.2	67.2	5,834.0	56.8	10,664.4	54.2	22,508.1	42.7
Basmati								
IRRI	1,176.4	2.3	12,270.0	10.8	10,400.8	14.4	21,628.3	14.9
Cotton								
Local								
Improved								
Sugarcane								
Maize								
Local								
Improved								

TABLE A.13 (continued)

Crop	Under 12.50 Acres		12.50-50.00 Acres		25.00-50.00 Acres		Over 50.00 Acres	
	Amount	(%)	Amount	(%)	Amount	(%)	Amount	(%)
	Larkana							
Wheat								
Local	1,015.9	7.9	1,397.8	5.4	2,847.7	7.9	5,717.4	7.6
Mexi-Pak								
Rice								
Local					1,644.0	0.3	1,369.5	0.4
Basmati	975.0	1.3			2,034.5	0.9	3,506.5	0.8
IRRI	4,946.1	71.0	11,323.0	58.1	20,626.2	57.1	35,996.2	47.9
Cotton								
Local								
Improved								
Sugarcane								
Maize								
Local								
Improved								
	Nawabshah							
Wheat								
Local	1,982.6	9.2	2,210.3	2.2	4,607.9	3.3	7,225.3	2.9
Mexi-Pak	4,841.4	11.2	8,947.3	8.7	13,106.9	7.9	24,601.8	6.2
Rice								
Local								
Basmati								
IRRI								
Cotton								
Local	5,882.9	10.2	3,846.3	2.8			42,365.0	2.1
Improved	7,984.3	37.1	9,299.4	18.1	20,016.4	26.3	38,822.9	23.5
Sugarcane	10,073.0	17.5	27,911.1	54.4	53,996.7	58.1	88,634.4	53.8
Maize								
Local							11,410.0	0.6
Improved								
	Hyderabad							
Wheat								
Local								
Mexi-Pak	1,199.5	21.3	2,020.8	16.1	2,389.9	12.0	8,736.0	16.0
Rice								
Local	557.0	1.0	703.5	0.6	1,206.5	0.7	1,501.7	0.5
Basmati								
IRRI	2,515.2	23.5	3,285.3	15.4	5,830.0	16.5	9,446.4	10.8
Cotton								
Local	1,422.6	5.3	5,085.0	2.4	2,641.5	1.6	9,072.3	3.1
Improved	1,839.8	17.2	4,685.9	24.2	6,958.1	19.6	25,859.0	26.6
Sugarcane	4,880.9	18.3	7,166.9	26.9	11,680.8	40.3	13,911.1	22.3
Maize								
Local	847.0	0.8	274.0	0.1				
Improved								

Note: Blank spaces indicate data not applicable.

Source: Compiled by the author.

TABLE A.14

Distribution of Average Net Farm Income per Acre, by Crop
(rupees)

Crop	Under 12.50 Acres	12.50-25.00 Acres	25.00-50.00 Acres	Over 50.00 Acres	Under 12.50 Acres	12.50-25.00 Acres	25.00-50.00 Acres	Over 50.00 Acres	Under 12.50 Acres	12.50-25.00 Acres	25.00-50.00 Acres	Over 50.00 Acres
	Jhelum				Gujranwala				Sahiwal			
Wheat												
Local	69.0	43.6	84.9	167.5	117.6	82.9	146.7	208.2	276.5	268.7	262.7	257.0
Mexi-Pak				22.1	213.9	286.7	293.1	326.1	313.9	403.6	432.2	436.3
Rice												
Local					445.0	286.5	280.9	348.0	306.9	298.7	277.7	239.3
Basmati					595.1	414.4	503.5	601.4	591.3	699.2	615.3	1,025.5
IRRI					374.1	271.8	306.4	338.7	473.0	427.6	454.4	493.3
Cotton												
Local						61.2	325.1		615.0	653.0	707.9	804.5
Improved									1,141.3	1,337.5	1,313.9	1,300.1
Sugarcane					1,424.1	1,091.2	930.0	508.1	1,191.1	1,106.1	1,255.5	1,125.2
Maize												
Local		436.9	404.0			32.4			150.0			
Improved									221.2	220.5	186.6	196.6
	Lyallpur				Rahimyar Khan				Jacobabad			
Wheat												
Local					79.5	113.2	76.6	135.4	184.4	186.9	137.9	234.5
Mexi-Pak	347.3	368.2	409.3	378.9	200.0	192.3	135.6	128.5		145.5	144.2	193.9
Rice												
Local	1,940.0	187.0	179.2			810.6	253.7	38.7	347.2	311.5	327.6	331.4
Basmati	363.3	441.8	561.5	401.9			1,010.6					
IRRI						811.6			588.2	557.7	480.0	436.9
Cotton												
Local					673.4	475.1	676.5	492.7				
Improved	636.3	769.6	989.5	1,271.9	711.6	813.3	756.4	715.3				
Sugarcane	1,073.8	1,205.9	1,645.5	1,976.0	1,156.3	1,251.9	1,426.3	1,317.8				
Maize												
Local	126.3	175.1			96.0	130.0	184.0	65.0				
Improved	420.2	508.2	534.1	436.2								

	Larkana				Nawabshah				Hyderabad			
Wheat												
Local	113.4	126.2	161.2	217.8	254.9	262.5	247.9	266.3				
Mexi-Pak					739.1	576.9	624.1	754.7	316.5	327.0	231.8	315.5
Rice												
Local			411.0	512.9					247.6	216.5	201.1	112.7
Basmati	487.5		678.2	779.2								
IRRI	634.1	587.0	580.9	626.0					433.7	305.6	323.9	317.0
Cotton												
Local					882.0	607.6		1,694.6	711.3	635.6	660.4	688.9
Improved					1,157.2	983.0	994.9	1,164.8	634.4	781.0	802.5	773.3
Sugarcane					2,664.8	2,098.6	2,216.6	2,287.3	2,000.4	2,389.0	2,038.5	1,918.8
Maize												
Local								1,141.0	847.0	274.0		
Improved												

Note: Blank spaces indicate data not applicable

Source: Compiled by the author.

TABLE A.15

Farm Income from Sale of Inputs and Outputs Other Than Crops
(rupees)

District/ Farm Size (acres)	Human Labor			Animal Labor			Farm Machinery		
	No. of Man-Days	Rate per Man-Day (Rs.)	Avg. Income (Rs.)	No. of Days	Rate per Day (Rs.)	Avg. Income (Rs.)	No. of Hours	Rate per Hour (Rs.)	Avg. Income (Rs.)
Jhelum									
Under 12.50	235.0	7.0	1,645.0						
12.50-25.00									
25.00-50.00									
Over 50.00									
Gujranwala									
Under 12.50									
12.50-25.00	6.0	10.0	60.0	6.0	10.0	60.0			
25.00-50.00							126.7	10.0	1,266.6
Over 50.00							115.0	10.0	1,150.0
Sahiwal									
Under 12.50	44.1	7.9	245.0	44.1	7.9	245.0			
12.50-25.00	12.2	8.3	168.3	12.2	8.3	168.3			
25.00-50.00	18.3	4.8	84.4	18.3	4.8	84.4	124.5	9.2	1,125.0
Over 50.00							166.7	9.4	1,540.7
Lyallpur									
Under 12.50						100.0			
12.50-25.00						300.0			
25.00-50.00									325.0
Over 50.00							213.2	8.5	2,541.2
Rahimyar Khan									
Under 12.50	332.0	4.8	1,546.5				400.0	20.0	800.0
12.50-25.00									
25.00-50.00							100.0	30.0	3,000.0
Over 50.00							1,150.0	15.0	11,500.0
Jacobabad									
Under 12.50	285.0	6.0	1,691.2	90.0	6.0	540.0			
12.50-25.00	235.0	6.5	1,555.0						
25.00-50.00	208.0	6.6	1,385.0						
Over 50.00	330.0	8.0	2,640.0				540.0	24.0	12,960.0
Larkana									
Under 12.50	211.0	8.3	1,742.5						
12.50-25.00	243.0	8.0	1,960.0						
25.00-50.00	243.0	8.5	2,062.5						
Over 50.00	244.0	8.8	2,194.2						
Nawabshah									
Under 12.50	262.0	7.2	1,902.1						
12.50-25.00	230.0	7.8	1,970.9						
25.00-50.00	201.0	7.8	1,565.4						
Over 50.00	204.0	8.0	1,554.0				330.0	30.0	9,900.0
Hyderabad									
Under 12.50	322.0	6.5	2,107.5						
12.50-25.00	373.0	5.5	2,187.1	100.0	30.0	3,000.0			
25.00-50.00	312.0	8.2	2,340.0						
Over 50.00	660.0	10.0	6,600.0				300.0	16.5	4,700.0

Note: Blank spaces indicate data not applicable.

Source: Compiled by the author.

Tubewell Water			Transport			Avg. Income from Other Sources (Rs.)	Avg. Income from All Inputs (Rs.)
No. of Hours	Rate per Hour (Rs.)	Avg. Income (Rs.)	No. of Miles	Rate per Mile (Rs.)	Avg. Income (Rs.)		
						4,122.5	2,586.8
						1,275.0	1,275.0
						2,130.0	2,240.0
						4,812.0	4,812.0
						2,010.0	2,010.0
						2,400.0	677.5
136.8	4.9	657.5	4,660.0	0.3	1,400.0		771.3
133.6	4.9	647.5	3,833.0	0.3	1,150.0	5,000.0	1,363.7
116.9	4.5	500.0				3,049.5	1,818.3
122.3	4.9	596.4				3,187.5	856.9
178.1	4.9	880.6	6,133.0	0.2	1,080.0	4,750.0	980.1
190.1	4.9	930.9	8,999.0	0.2	1,787.9		1,473.7
						1,740.0	1,505.7
						2,484.0	1,860.0
		1,000.0			750.0	4,688.4	5,572.0
121.0	5.0	2,309.0	4,000.0	0.3	1,190.5	7,126.2	3,235.1
50.0	5.0	250.0				1,300.0	1,201.8
100.0	5.0	500.0				1,600.0	1,350.0
140.0	5.4	740.0				13,100.0	2,975.0
						3,974.5	5,468.1
						3,491.2	2,575.9
						2,774.5	2,338.3
						3,316.3	3,643.9
						5,790.5	5,773.9
						2,056.7	1,868.1
						2,680.0	2,375.6
						3,330.0	2,411.5
						4,336.2	2,830.2
						1,825.0	1,888.1
						2,425.0	1,787.7
						2,100.0	1,579.1
						2,820.0	2,280.0
						1,337.5	1,775.7
						1,600.0	2,137.8
			150.0	3.0	450.0	2,212.5	2,415.0
150.0	12.0	1,050.0				16,097.5	7,986.6

TABLE A.16

Distribution of Farm Indebtedness

District	Farmers in Debt (%)	Average Size of Debt (Rs.)	Purpose of Debt: Consumption (%)	Purpose of Debt: Production (%)	Purpose of Debt: Other (%)	Distribution of Debt by Source — Institutional: ADBP (%)	Institutional: Cooperatives (%)	Institutional: Taccavi (%)	Noninstitutional: Commission Agents (%)	Noninstitutional: Friends and Relatives (%)	Noninstitutional: Moneylenders (%)	Noninstitutional: Other (%)
Jhelum												
Landless	25.0	1,150.0	100.0							100.0		
Under 12.50	75.0	801.3		93.6		6.4	2.1			91.5		
12.50-25.00	42.9	1,733.3		84.6	15.4					100.0		
25.00-50.00	100.0	2,666.7			100.0	25.0				75.0		
Over 50.00	50.0	350.0		100.0		100.0						
Gujrunwala												
Landless												
Under 12.50	42.9	250.0	46.7	40.0	13.3					53.3		46.7
12.50-25.00												
25.00-50.00	7.1	15,500.0		96.8	3.2	96.8				3.2		
Over 50.00	58.3	11,611.1		100.0		95.7			3.4	1.0		
Sahiwal												
Landless												
Under 12.50	67.4	387.5	30.2	10.2	58.2				55.3	29.2		22.9
12.50-25.00	44.9	687.0	14.8	67.4	17.8	16.2			31.3	49.3		3.2
25.00-50.00	30.2	9,652.9	1.1	98.1	0.8	64.0		3.0	8.0	0.5		0.2
Over 50.00	41.4	32,394.7		100.0		99.9			0.1			
Lyallpur												
Landless												
Under 12.50	76.3	231.7		68.3	31.7					77.7		22.3
12.50-25.00	71.1	460.6		71.7	28.3		6.6		13.2	53.9		24.3
25.00-50.00	68.4	1,223.1		72.3	27.7	41.9	6.3		14.7	24.1		14.1
Over 50.00	76.5	18,000.0		98.2	0.3	77.1	4.4		11.3	7.2		

Rahimyar Khan												
Landless	45.0	561.1	60.4	39.6						81.2		18.8
Under 12.50	35.0	1,585.7	1.8	98.2			6.3			93.7		
12.50-25.00	40.0	3,995.6	40.7	59.3		72.0	4.7			15.6		7.7
25.00-50.00	50.0	7,145.5	3.8	96.2		62.5	1.3		3.8	10.2		22.3
Over 50.00	46.7	16,500.0	8.7	91.3		56.7				43.3		
Jacobabad												
Landless	33.3	1,625.0		100.0						7.7	92.3	
Under 12.50	36.4	1,450.0	22.4	77.6					70.7		29.3	
12.50-25.00	41.7	1,520.0	34.2	52.6	13.2				79.0		21.1	
25.00-50.00	41.7	2,300.0	21.7		78.3				26.1	4.4	69.6	
Over 50.00	100.0	4,725.0	2.1	63.5	34.4			17.6	65.6	2.6	14.1	
Larkana												
Landless	33.3	2,600.0	7.7	92.3							100.0	
Under 12.50	41.7	1,500.0	13.3	40.0	46.7				53.3		46.7	
12.50-25.00	58.3	2,814.3		84.8	10.2				79.7		20.3	
25.00-50.00	25.0	2,833.3		100.0					100.0			
Over 50.00	50.0	4,166.7	4.0	60.0	36.0			20.0	64.0		16.0	
Nawabshah												
Landless	28.6	1,625.0	7.7	92.3					38.5		61.5	
Under 12.50	21.4	1,500.0		100.0					100.0			
12.50-25.00	21.4	2,216.7	30.1	69.9					100.0			
25.00-50.00	28.6	2,537.5		80.3	19.7				100.0			
Over 50.00	35.7	5,800.0		17.2	82.8	51.7			41.4		6.9	
Hyderabad												
Landless	36.4	593.8	70.5	29.5						100.0		
Under 12.50	86.4	1,347.4	92.2	7.8				11.7			88.3	
12.50-25.00	50.0	1,680.0	94.6	5.4				21.6		54.1	24.2	
25.00-50.00	40.0	11,062.5	7.3	92.7		84.7				7.9	7.3	
Over 50.00	30.0	31,333.3	1.6	98.4		98.4						1.6

Notes: ADBP is Agricultural Development Bank of Pakistan.
Taccavi is a government loan to farmers in distress or affected by a natural disaster.
Blank spaces indicate data not applicable.

Source: Compiled by the author.

TABLE A.17

Ratio of Average Value of Output per Acre to Average Variable Cost per Acre

District/Farm Size	All Crops	Mexi-Pak Wheat	IRRI Rice
Jhelum			
Under 12.50	4.14		
12.50-25.00	6.68		
25.00-50.00	4.98		
Over 50.00	2.51		
Gujranwala			
Under 12.50	5.50	3.51	4.86
12.50-25.00	4.90	3.56	3.58
25.00-50.00	4.06	3.48	3.74
Over 50.00	3.29	3.08	3.20
Sahiwal			
Under 12.50	5.56	3.58	4.72
12.50-25.00	6.29	4.17	3.71
25.00-50.00	6.33	4.14	4.33
Over 50.00	6.52	3.83	4.00
Lyallpur			
Under 12.50	5.70	4.19	
12.50-25.00	5.70	4.26	
25.00-50.00	5.31	3.68	
Over 50.00	4.91	2.74	
Rahimyar Khan			
Under 12.50	4.59	2.40	
12.50-25.00	4.99	2.35	
25.00-50.00	4.32	1.84	
Over 50.00	3.03	1.52	
Jacobabad			
Under 12.50	8.14		9.19
12.50-25.00	7.12	6.11	7.46
25.00-50.00	8.68	6.36	8.81
Over 50.0	6.56	6.04	5.61
Larkana			
Under 12.50	6.68		6.45
12.50-25.00	7.58		6.06
25.00-50.00	7.77		6.46
Over 50.00	9.50		7.02
Nawabshah			
Under 12.50	9.18	6.20	
12.50-25.00	4.14	5.41	
25.00-50.00	4.92	5.85	
Over 50.00	5.83	6.24	
Hyderabad			
Under 12.50	6.74	5.08	8.24
12.50-25.00	6.05	4.23	5.44
25.00-50.00	5.06	3.30	4.28
Over 50.00	5.35	3.28	4.44

Note: Blank spaces indicate data not applicable.

Source: Compiled by the author.

INDEX

ABOUT THE AUTHOR

MAHMOOD HASAN KHAN is Associate Professor in the Department of Economics and Commerce at Simon Fraser University, British Columbia, Canada. He was an OECD consultant to the Middle East Technical University in Ankara, Turkey, during 1970. He has received a research grant and leave fellowship from the Canada Council for his work on the "Green Revolution" in Pakistan.

Dr. Khan has published widely in the area of economic development and agricultural economics. His first book, published in Holland, was entitled *The Role of Agriculture in Economic Development: A Case Study of Pakistan*. He has contributed articles and book reviews to *Quarterly Journal of Economics*, *Economia Internazionale*, *Journal of Agricultural Economics*, and *Pacific Affairs*. He has written chapters in two books and has published articles in *Pakistan Economist*.

Dr. Khan holds a B.S. (Agric.) and M.A. from the University of Sind in Pakistan, and an M.Soc.Sc. and Ph.D. from the Agricultural University, Wageningen, Holland.

RELATED TITLES

Published by
Praeger Special Studies

FOOD, POPULATION, AND EMPLOYMENT:
The Impact of the Green Revolution
edited by Thomas T. Poleman
and Donald K. Freebairn

AFRICAN FARMERS: LABOR USE IN THE DEVELOPMENT OF SMALLHOLDER AGRICULTURE
John H. Cleave

EDUCATION, MANPOWER, AND DEVELOPMENT IN SOUTH AND SOUTHEAST ASIA
Muhammad Shamsul Huq

AFGHANISTAN IN THE 1970s
edited by Louis Dupree
and Linette Albert

SOUTH ASIAN CRISIS: INDIA, PAKISTAN, AND BANGLADESH
A Political and Historical Analysis of the 1971 War
Robert Jackson